KEY CONCEPTS IN
# URBAN
# GEOGRAPHY

**KEY CONCEPTS**
IN HUMAN GEOGRAPHY

The *Key Concepts in Human Geography* series is intended to provide a set of companion texts for the core fields of the discipline. To date, students and academics have been relatively poorly served with regards to detailed discussions of the key *concepts* that geographers use to think about and understand the world. Dictionary entries are usually terse and restricted in their depth of explanation. Student textbooks tend to provide broad overviews of particular topics or the philosophy of Human Geography, but rarely providing a detailed overview of particular concepts, their premises, development over time and empirical use. Research monographs most often focus on particular issues and a limited number of concepts at a very advanced level, so do not offer an expansive and accessible overview of the variety of concepts in use within a subdiscipline.

The *Key Concepts in Human Geography* series seeks to fill this gap, providing detailed description and discussion of the concepts that are at the heart of theoretical and empirical research in contemporary Human Geography. Each book consists of an introductory chapter that outlines the major conceptual developments over time along with approximately twenty-five entries on the core concepts that constitute the theoretical toolkit of geographers working within a specific subdiscipline. Each entry provides a detailed explanation of the concept, outlining contested definitions and approaches, the evolution of how the concept has been used to understand particular geographic phenomenon, and suggested further reading. In so doing, each book constitutes an invaluable companion guide to geographers grappling with how to research, understand and explain the world we inhabit.

Rob Kitchin
Series Editor

# KEY CONCEPTS IN
# URBAN GEOGRAPHY

ALAN LATHAM
DEREK McCORMACK
KIM McNAMARA
DONALD McNEILL

Los Angeles • London • New Delhi • Singapore • Washington DC

SAGE Publications Ltd
1 Oliver's Yard
55 City Road
London EC1Y 1SP

SAGE Publications Inc.
2455 Teller Road
Thousand Oaks, California 91320

SAGE Publications India Pvt Ltd
B 1/I 1 Mohan Cooperative Industrial Area
Mathura Road, Post Bag 7
New Delhi 110 044

SAGE Publications Asia-Pacific Pte Ltd
33 Pekin Street #02-01
Far East Square
Singapore 048763

**Library of Congress Control Number 2008922616**

**British Library Cataloguing in Publication data**

A catalogue record for this book is available from the British Library

ISBN 978-1-4129-3041-3
ISBN 978-1-4129-3042-0 (pbk)

Typeset by C&M Digitals (P) Ltd, Chennai, India
Printed by The Cromwell Press Ltd, Trowbridge, Wiltshire
Printed on paper from sustainable resources

# CONTENTS

# ACKNOWLEDGEMENTS

We would like to thank Rob Kitchin, Robert Rojek, Sarah-Jayne Boyd, Wendy Scott and the production staff at SAGE for all their help in moving the project from commissioning to completion. Derek would like to thank Brendan Bartley and Paul Knox for helping him think about urban geography, and Andrea for her patience. Alan would like to thank Rita and Tom for their love and support. Kim would like to thank her family, and especially Mia and Mikaela. Donald would like to thank Carol for her love.

Alan, Derek and Donald would like to show their appreciation of the guidance of Steven Pinch. Many of the ideas contained in their entries were developed while at the University of Southampton, and Steven's counsel and enthusiasm was invaluable. Kim is grateful for the advice of Nick Couldry.

**Photo acknowledgements**

All photos are copyright of Kim McNamara, except that of 2.3.1 and 5.1.2 (© Donald McNeill).

**Diagram acknowledgements**

1.1.2: Park, R., Burgess, E. and McKenzie, R. (1925/1967) *The City*, Chicago: University of Chicago, Chart II. Urban areas. p. 55.

3.1.1: 'Generalisations of internal structure of cities', Fig. 5, p. 13 in Harris, C. and Ullman, E. (1945) 'The nature of cities', *The Annals of the American Academy of Political and Social Science*, 242: 7–17.

3.1.2: Latham, A. (2003a) 'Research, performance, and doing human geography: some reflections on the diary-photo diary-interview method', *Environment and Planning A*, 35: 1993–2017. (The image is from page 2010.)

# Acknowledgements

4.1: 'The ecology of fear' from *Ecology of Fear: Los Angeles and the Imagination of Disaster*, by Mike Davis. Copyright 1998 by Mike Davis. Reprinted by permission of Henry Holt and Company, LLC.

# INTRODUCTION

It is easy to list the things that seem to belong to cities. Audacious skyscrapers. Bustling sidewalks. The hum of traffic. Elegant department stores. Subways and flyovers. Ethnic restaurants and street cafes. Squares, parks and piazzas teeming with people. Shopping malls and industrial parks. Giant sports stadiums. The cries of street hawkers. Trolley cars and double-decker buses. Outrageously dressed people. Street markets. Slums and tenements. Beggars and drunks. Soot stained buildings. Smog. Foul tasting tap water. We should not forget exhibition halls and conference centres, art galleries and cultural quarters, Staffordshire bull terriers and corner gangs, warehouse conversions and new media start-ups. Nor monuments and memorials, skid rows and strip malls. The list goes on and on. It is also pretty easy to list the kinds of adjectives that convey some sense of what cities are about. Fast, loud, noisy, brash. Diverse, cosmopolitan, superficial. Unnatural, inhuman, anonymous. Risky, sexy, dazzling, dangerous. All these words convey something about how cities are experienced, and how cities are understood. But it is hard to define what makes a city a city, and what makes life urban. For example:

1

> The city is everywhere and in everything. If the urbanized world now is a chain of metropolitan areas connected by places/corridors of communication (airports and airways, stations and railways, parking lots and motorways, teleports and information highways) then what is not the urban? Is it the town, the village, the countryside? Maybe, but only to a limited degree. The footprints of the city are all over these places, in the form of city commuters, tourists, teleworking, the media, and the urbanization of lifestyles. The traditional divide between the city and the countryside has been perforated. (Amin and Thrift, 2002: 1)

So, we are faced with a problem. While many of the themes that excite urban geographers may at first sight be associated with the dense melting pots of the world's most exciting places, we can also think of a list of things that are a product of urban society, but not necessarily of the *city*: bento boxes, golf balls, soap operas, barbecues, shipping containers, credit cards, filing cabinets, petrol pumps, microwave ovens, grass verges at the sides of roads, birthday cards, rubbish dumps, melting ice floes. The footprints of the city are, indeed, as much in the urbanisation of lifestyles as in any spatial form.

# Defining cities, defining the Urban

So, how do we begin to define our terrain of interest? Definitional questions are quite important to consider, as we may be using cities as a 'spatial fix' or 'fetish', a means of projecting social relations onto a defined territory in order to make them more understandable (Beauregard, 1993; Pile, 1999). There tends to be a slippage between talking about cities, and talking about the urban, both in academic texts and in popular usage. The quotation from Amin and Thrift above draws on the useful idea of a 'footprint', where even the countryside becomes urbanised through the industrial-scale manufacture of food, or the commuting urban dweller moving into a village cottage, or the television showing the mean streets of an American crime series in the living room of a remote village. Certainly, the creation of nationwide networks of motorways, railways and air routes brings even remote areas within an urban time-space, thus lessening a distinction between the country and the city.

However, most people would agree that an important – perhaps defining – element of cities is their human *density*. This has a legal dimension. In medieval Europe a city was any town that had been empowered by the local nobility to build a city wall. In the United States, 'city' refers to an incorporated urban area with powers of self-government. Cities like Los Angeles or Chicago, legally at least, are made up of scores (in some cases hundreds) of incorporated cities – they are quite literally cities of cities (Fug, 1999; D. Massey, 2005). And these cities have only the loosest relationship to population numbers. The smallest city in the Los Angeles metropolitan area, Industry, has a residential population of just 777. Nonetheless, attracted by the absence of local business taxes, Industry is home to over 2000 businesses and 80,000 jobs (see Davis, 1990). These raw numbers do point to the fact that sheer size of population (and with it settlement density) in some sense structures the dynamics of a city, that it helps give a city its *city-ness*, if you like. They also point to the fact that a sense of political incorporation, some sense of shared interdependency, lies at the centre of what cities are about. And, the example of Industry points to the way that cities are not just about who lives within their boundaries, but also all the activities and other entities (companies, workers, factories, and so forth) that populate them. Here, we can see the importance of cities as sites of interaction and association, involving new forms of social relationships (between children in school playgrounds, car drivers in complex traffic routes, business people stitching together transactions, friends or co-believers congregating to celebrate football or religious beliefs).

It is important to also consider the geography of knowledge production in urban geography. One of the great temptations in all work on cities – which by its very nature tends often to be case study based and place bound – is to claim that the city being studied is definitive of the thing being studied. The book edited by Robert Park and Ernest Burgess (1925) that defined the Chicago School of urban research in the 1920s went by the rather generic title *The City: Suggestions for Investigation of Human Behaviour in the Urban Environment* despite the fact that it is almost entirely based on research undertaken in Chicago. More recently, a group of urban studies scholars have pushed forward the claim that Los Angeles is the best place to witness the future form of metropolitan life. As Ed Soja (1989: 191) claimed:

> What better place [than Los Angeles] can there be to illustrate and synthesize the dynamics of capitalist spatialization? In so many ways, Los Angeles is the place where 'it all comes together', to borrow the immodest slogan of the *Los Angeles Times*. Being more inventive, one might call the sprawling urban region defined by a sixty-mile circle around the centre of the City of Los Angeles a *prototopos*, a paradigmatic place; or pushing inventiveness still further, a *mesocosm*, an ordered world in which the micro and the macro, the idiographic and the nomothetic, the concrete and the abstract, can be seen simultaneously in an articulated and interactive combination.

3

The difficulty with these kinds of arguments is threefold. First, why choose Los Angeles, or Chicago? Why not New York or London, Detroit or Warsaw, Osaka or Houston, Miami or Kolkata, or Lagos, or any other interesting looking city? Second, the framework used in the analysis tends to homogenize cities, seeking to apply clinical laboratory conditions to an incredibly complex collection of social formations. In the words of Ash Amin and Steven Graham (1997: 417) 'the exception [...] becomes the norm, applicable to a vast majority of what might be called "unexceptional" cities'. So behind the glamorous representations of Los Angeles, New York, London and Shanghai, the urban condition is being lived out in fairly unremarkable urban places. For example, a recent book entitled *Small Cities* addresses 'the woeful neglect of the small city in the literature on urban studies' and seeks to outline 'appropriate ways to understand what small cities are, what smallness and bigness mean, how small cities fit or don't fit into "the new urban order", or what their fortunes or fates might be' (Bell and Jayne, 2006: 2). Relatedly, there is the notion of the urban as an ever-increasing thing: 'the discourses of cities ... have tended to follow the logic that cities should be big things, either amazing or terrifying in their bigness, but big nonetheless. The very idea of cities is to be big and to get bigger: shrinkage, even stasis, is a sign of failure.' (Bell and Jayne, 2006: 5). Third, such cities can be expressive of a longstanding divide within geography, where the study of cities in the so-called 'Third World' is often undertaken within the sub-discipline of development studies, rather than urban geography. For example, it has been argued:

> the limited applicability of different accounts of (western) cities conventionally remains unstated, even if it is implicit in the context ... This is much less the case for writing about cities outside the West, where explicit naming of the region or cities covered highlights the implicit universalist assumptions underpinning the often unremarked localness of much writing on western cities. (Robinson, 2006: 543)

This is not to say that such claims are not without their uses. A term like the Chicago School, or the Los Angeles School, becomes shorthand for a series of distinctive arguments about how cities are structured. And the hyperbolic claims for paradigmatic status ('new kind of city', 'biggest', 'fastest growing', 'most dynamic', 'where it all comes together') can, as Michael Dear (2003: 202) argues, help jolt readers out of received ways of thinking and 'encourage new ways of seeing' cities. However, focusing on the exceptional and the remarkable can cause one to miss the ordinary, the small variations that actually make a particular urban space, or city, distinctive. Instead, we suggest that it is important to think about cities and the urban 'as the co-presence of multiple spaces, multiple times and multiple webs of relations, tying local sites, subjects and fragments into globalising networks of economic, social and cultural change' (Amin and Graham, 1997: 417–8).

So, the purpose of *Key Concepts in Urban Geography* is to present a series of ways of thinking about what makes cities what they are. The 20 chapters that make up the book offer a series of routes into the complexity, the heterogeneity, the ambiguity and the dynamism of urban life. We want to consider how cities are not just defined by propinquity (by things being physically close together) but also by a whole range of long distance relationships – connections to the countryside around it, to other cities, to other countries.

# Urban geography as a sub-discipline

Urban geography is a relatively youthful sub-discipline. Early work focused on the ways that climate and local geographical conditions had shaped the development of individual urban centres. This certainly produced some interesting work. Jean Brunhes' (1920) *Human Geography*, for example, showed in great detail how the architecture and form of particular settlements could be related to the surrounding region's topography, climate and geology. But there was only so much to be said about the importance of the immediate physical environment to things like settlement location, or a city's internal morphology. It is only in the past 50 years that urban geography has emerged as a recognisable entity either within geography or, more broadly, within social scientific research on cities and urbanisation. As recently as the end of the 1950s, urban geography was defined by just a handful of textbooks and monographs (see Dickinson, 1947; Taylor, 1949; Mayer and Kohn, 1959) and, in North America and the United Kingdom at least, it was rare for university geography departments to offer undergraduate courses focusing exclusively on urban issues. Prior to the 1950s geography in general was preoccupied by two principal themes: first, that of regional distinctiveness and second, on the relationship between humans and the physical environment. Geography was thus essentially an idiographic discipline (a discipline concerned with singular empirical cases) closely aligned with the humanities, and a discipline where the division between human and physical geography was blurry (humans were after all part of the physical environment).

This situation altered dramatically in the early 1960s, as urban geography emerged as a quantitatively based 'spatial science', defined by three central principles. First, it was convinced that the aim of geography should be to find general laws that defined the formation of geographical relationships. Second, to discover these relationships geography had to develop rigorous and empirically testable theories. And third, geography had to adopt established scientific techniques such as statistical analysis, hypothesis testing, or mathematical modelling, to properly test these theories. Thus, to be scientific, data needed to be measurable and countable. And to be rigorous, data needed to be scientifically tested against some theory about how the world worked. While these principles may appear straightforward enough and even rather quaint and a little naive to contemporary ears, they represented a revolution within the conservative world of early 1950s geography.

5

hidden (literally and metaphorically) from our attention. And it has also invigorated debates in the field of **architecture**, both in terms of the networks of relations that come together to make a building happen and also through an understanding of how buildings are used, both in the everyday sense of home and housing and in ceremonial, identity-marking institutions such as state parliaments or museums.

## (3) Envisioning and Experience

A major critique of quantitative urban geography was that its under-standing of human action was too narrow and unnecessarily constric-tive. These geographers were accused of having overlooked the social and political dimensions of the urban forms that they studied, or even excluded an interest in human *experience*. By seeing human actors as driven by neo-classical presupposition of strict economic rationality and, further, by suggesting that the only dimensions of human action that could be studied scientifically were those that could be rigorously measured, quantitative urban geography was blind to what it is that makes humans human. Emotion, memory, our ability to form mean-ingful attachments with each other, our capacity for wonder, all had no place in quantitative urban geography. A key issue, however, was an unwillingness to accept that how they envisioned the urban was a very loaded decision. The **diagrams** used in these urban geographies are important to consider, as they both imply a scientific form of 'master-ing' urban space, as well as reducing human behaviour to a series of lines drawn on maps.

In contrast, humanistic geographers such as Edward Relph, Yi Fu Tuan and David Ley suggested a quite different way of doing urban geography, one that in the words of Anne Buttimer (1978: 74) 'allows for emotion as well as thinking, passion as well as reason.' Feminist geographers like Susan Hansen, Linda McDowell, Gillian Rose and Doreen Massey too demanded a different kind of urban geography. They pointed out not only that women (and indeed children) experienced urban space in very different ways to men, but also that urban space was *gendered* in all sorts of complicated and rarely acknowledged ways (see Ley and Samuels, 1978; Ley, 1983; Women and Geography Study Group of the IBG, 1984; Rose, 1993). In her book *Visual Methodologies* (2007), Gillian Rose set out a comprehensive framework for thinking through how visual methods used by social scientists could unveil some of these experiences.

These literatures underpinned the message that cities are more than physical structures: they are also sites of meaning and experience. Consider, for example, the way in which 'fast' technologies such as railways – and then airplanes – have transformed our understanding of cities. Theorists such as Schivelbusch (1986) and Schwarzer (2004) have argued that visuality is central to the modern urban experience:

> Panoramic vision turns the view of the city into a sequence of disembodied and abstracted forms. Schivelbusch realizes that since rail passengers perceive specific objects poorly, they tend not to look closely or carefully. Speed anaesthetizes vision. Sight becomes absentminded. Instead of observing a building's form, rail passengers see odd features in the shifting juxtapositions brought about by the train's velocity and their own haphazard concentration. A new type of building is seen. This is not the building carefully designed by the architect, but instead a building interconnected with other buildings, other objects, and other images in the mind. (Schwarzer, 2004: 54)

Railway or car journeys are thus interesting ways of thinking through how the urban is constituted, as it suggests that many urban dwellers switch into absent-mindedness when travelling through complex urban landscapes. By contrast, there is also a tendency to associate cities through *visual* metaphors, associating them with their distinguishing 'trademarks' such as Sydney's harbour, bridge and opera house; London's palaces and Victorian buildings; Rome's monuments and ruins or New York's skyline of high-rise towers. Moreover, they are often dominated by complex forms of visuality in how we make sense of urban space, most frequently expressed through the medium of **photography**, which has been important in the rise of cinematic productions and still images alike.

11

To consider this further, the next entry explores the importance of the **body** as an important site of research in itself, not least given the growing obsession with physical appearance, body shape and fashion, which have all been important sectors in the contemporary western economy. The body can be seen as a site of social action in its own right, as much as being 'placed' in the landscape. It is also the mechanism by which humans experience urbanity through sensory organs, and allow for an understanding of how emotions – which are usually contrasted with an idealisation of city dwellers as taking 'rational' decisions about housing, commuting and so on – arguably rule urban life.

We then focus on how urban experience can be understood via the concept of **virtuality**, not just in terms of the digital, but in terms of an experience that points to an imaginative and future-oriented sense of experience. While often associated with the rise of 'cybercities', virtuality has a more complex set of meanings, which includes the practice of imagining urban space, the use of technology to simulate 'real' spaces and a complex set of transactions in time, most notably in financial trading.

Finally, we consider a concept – **surveillance** – that allows us to get a handle on how various forms of governance infiltrate many of the most everyday of urban practices and routines. The increasing sophistication of camera technology has allowed public spaces to be increasingly monitored visually. Yet surveillance means much more than this, as corporations seek to scan consumer behaviour in the search for greater market sensitivity, governments construct databases to watch over who is part of the national community and the military adopts cartographic techniques in order to enhance their operations (often against civilian targets).

## (4) Social and Political Organisation

As noted above, the power of social and political critique was given an important stimulus by developments in Marxian geography. As with the humanistic and feminist critiques of quantitative urban geography, the emergence of a Marxian (or political economic) urban geography brought with it a whole host of new thinkers, and intellectual traditions. Marxian geographers were inspired not by established geographers such as Halford Mackinder and Walter Christaller, but to Engels with his writings on Victorian Manchester, Lenin and Luxemburg with their writings on the geographical expansion and intensification of capitalism, Lefebvre with his theses about the urbanisation of capitalism, and Marx himself. Marxian urban geography was also concerned with an entirely new set of research themes (trade unions, political activists, ideological fields, the dynamics of capitalist accumulation) and a whole new range of empirical concerns, not least of which was understanding how the economic structures of a city were intertwined with its political institutions.

In 1973, David Harvey published *Social Justice and the City*, now seen as a turning point in the use of Marxian concepts in urban geography. Starting out as a politically liberal meditation on the relationship between cities and social justice, Harvey came to the conclusion that liberal – that is to say mainstream – social science was incapable of understanding the underlying causes of the many inequalities and social injustices that structured the experience of the modern city. His arguments inspired a

12

wide range of social geographic research. This is not to deny the significance of quantitative, statistical measures of **segregation** as captured by census data and other socio-economic surveys, but rather to take a more reflective, integrated approach to considering the categorisation of such data. In this context, the importance of recognising 'difference' has been an important motivation for geographers working at the intersection of post-colonialism and urban geography and planning (e.g., Fincher and Jacobs, 1998).

Harvey was not alone in his call for a 'radical' (that is to say anti-establishment), Marxian inspired, urban geography. Since 1969 the journal *Antipode* had been publishing Marxian inspired (along with feminist and other) critiques of quantitative geography. But *Social Justice and the City* acted as a catalyst in redefining what a radical urban geography would be about – not least because it asserted, first, that geography was absolutely central to the dynamics of the capitalist system and, second, that cities in particular were key sites for the realisation of surplus value – that they were money-machines. In the following years, a multitude of geographical scholars have extended the scope of Marxian urban geography. Indeed, if quantitative geography defined the dominant intellectual trajectory within urban geography from the mid-1950s into the 1970s, political, economic and Marxian approaches dominated urban geography through much of the 1980s and early 1990s and helped shape the dominant concepts used in explaining the form of **urban politics**.

However, Marxian geography was but one of a series of streams that entered urban geography via sociology. An important theme of urban studies throughout the twentieth century has been that of **community**. This entry traces out the diverse set of ideas that have underpinned this slippery term and argues that the literature has moved from an idea of 'community lost' (where knowing one's neighbour, for example, is an important theme of successful neighbourhoods) to 'community saved', where vibrant social relations are found either in rejuvenated – and perhaps gentrified – inner-city neighbourhoods, or else in a more distanciated form, via the internet, or even in more complex forms of interaction between human and object, where community is redefined.

## (5) Sites and Practices

The final section allows us to consider some of the topographies of contemporary urban life. Again, Marxian thinkers have been influential

13

here, particularly those of the Frankfurt School, associated with the writings of Theodore Adorno and Max Horkheimer. Their conceptualisation of a 'culture industry' that pacified the masses with consumer goods and mass media entertainment was influential to many scholars, eager to explain why social groups in post-war societies seemed to be deradicalised and passive political bystanders, rather than active opponents of established social norms. Such work tended to ignore the desire of audiences to actively choose the consumer goods and entertainment choices that would make up their own modes of urbanised living. Recent work in geography, sociology, anthropology and media studies has tackled this issue head-on, particularly in terms of the sites of **consumption** in cities, from the spectacular to the mundane. In many ways, the mass production of new goods was dependent upon an urbanised society, 'where knowledge of consumption was essentially practical knowledge, not acquired through instruction or advertising, but from the experience of participating in activities in the dense interaction and information networks of urban life' (Glennie and Thrift, 1992: 430).

An understanding of the **media** is important in terms of making sense of how cities are represented, and in the ways in which the urban is produced, distributed and consumed. This is an emerging area of urban studies, not one that geographers have contributed a whole lot to. Yet in media studies and sociology, attempts are being made to conceptualise how media is at the same time a material practice (revolving around television studios, production companies and satellite infrastructure), a textual representation (in terms of its distribution of symbolic collections of words and images that are packaged, sold and consumed), and a relational process, in that places are linked together by media practices. This runs between of extremes, from the state-centred propaganda machines that dominate programming in China, to the completely decentralised chatrooms of the 'blogosphere'.

Such forms of social interaction pose important questions about the nature of being 'public' in the city. The entry on **public space** described some of the debates around everyday life in cities, particularly the apparent decline of political expression in public. It is suggested that – contrary to the orthodox view of there being a decline in sociality in cities – there are many new and vibrant forms of urban sociality.

Being *seen* in public is an important aspect of this, and multiple publics often seek out symbolic sites of commemoration or collective identity to express particular world-views. Recent events such as 9/11 have added to the sense of immediacy felt by geographers and others in explaining and understanding such events. Our final entry concerns the places and practices of **commemoration** within urban space, given their importance both as a form of representation of a particularly admired historical figure, a source of contest and conflict (as was the case in post-1989 East and Central Europe) or as a focus of collective or individual displays of grief, anger or joy.

# 1 Location and Movement

# 1.1 CENTRALITY

Centrality has always been a concept that has fascinated urban geographers. Just as metropolitan city-regions developed in the post-war period in ways unrecognisable from the early industrial city, so urban geography developed as a discipline (Wheeler, 2005). The arrival of the early mainframe computers in the 1960s gave birth to the 'science' of spatial analysis, where the powerful statistical tools afforded by the new technology saw a growing passion for quantifying everything about urban life – demographic trends, migration movements, housing markets, journey-to-work trips and so on (Barnes, 2003). In its earliest manifestations, such as the Central Place Theory of Walter Christaller, or the concentric-zonal, sectoral and multiple nuclei models of Ernest Burgess, Homer Hoyt and Harris and Ullman respectively, who each argued that cities had a discernible internal structure, (see figures 1.1.2 and 3.1.1) usually based upon access to markets and a tendency for activities to cluster in central places.

The power of most cities has tended to emerge due to locational advantage of some sort. From the simplest forms of exchange, when peasant farmers literally brought their produce from the fields into the densest point of interaction – giving us market towns – the significance of central places to surrounding territories began to be asserted. As cities grew in complexity, the major civic institutions, from seats of government to religious buildings, would also come to dominate these points of convergence. Large central squares or open spaces reflected the importance of collective gatherings in city life, such as Tiananmen Square in Beijing, the Zócalo in Mexico City, the Piazza Navona in Rome and Trafalgar Square in London. These manifestations of bustling centrality appeared to obey a gravitational pull. In certain extreme cases, such as Madrid and Brasilia, new capital cities were located at the central point of national territory for the most rational form of centralised governance. However, these examples suggest that urban structure was more often than not a process driven by social class and governmental power (especially in colonial cities, where 'enlightened' town planning created monumental squares, grandiose government buildings and wide boulevards) than by strict commercial criteria.

Nonetheless, for most modern cities the exemplar of centrality is its central business district. As the likes of Fogelson (2001) have described, American 'downtowns' have always been highly contested spaces, given the land values associated with being at the centre of transactional density (transport hubs, retail markets, office buildings, theatres, etc.). Groups of landowners began to tussle over building heights, subways and streetcars from the earliest periods of the modern city. Business groups have long been aware of the reduced transaction costs involved in concentrating business activities, most notably the minimisation of travel time when moving between clients. By the late nineteenth century, technological change had hastened the development of skyscrapers in business districts, much to the chagrin of those sections of the middle classes whose views were obscured by the new high-rises, as well as existing landowners whose property values were undermined by the sudden onslaught of new built space.

**Figure 1.1** 'Downtown Chicago has been a laboratory for urban geographers'

The cause of such conflicts has a simple economic logic:

All functions occupying space within the CBD have in common the need for centrality and their ability to purchase accessible locations. Within the CBD there are diversities which are revealed by the distinctive functional

districts and by the individual locational qualities of specific functions ...
The retail trade quarter is often referred to as the node and is usually on the
most central space; the office quarter is well-marked and may have sub-
sections such as financial or legal districts. Besides these horizontal divisions,
there are distinctive vertical variations in the distribution of functions; the
ground-floors of multi-storey buildings are occupied by activities with the
greatest centrality needs. (Herbert, 1972: 90)

However, with increased dispersal and stretching out of economic
relations, renting or owning an office or shop in the historically central
location began to decline in importance. The question of 'central to
what' became more important. In the 1930s and 1940s United States,
for example, inner areas became challenged by suburban business
districts, made possible by 'streetcar suburbs' and a burgeoning
mortgage industry. The sight of boarded up shops and theatres became
commonplace and a growing fear of downtown became internalised
within social discourse. As Robert Beauregard argues in his important
book *Voices of Decline* (1993):

The city is used rhetorically to frame the precariousness of existence in a
modern world, with urban decline serving as a symbolic cover for more wide-
ranging fears and anxieties. In this role, urban decline discursively precedes
the city's deteriorating conditions and its bleak future. The genesis of the discourse is not the entrenchment of poverty, the spreading of blight, the fiscal
weakness of city governments, and the ghettoization of African-Americans,
but society's deepening contradictions. To this extent, the discourse functions to site decline in the cities. It provides a spatial fix for our more general-
ized insecurities and complaints, thereby minimizing their evolution into a
more radical critique of American society. (Beauregard, 1993: 6)

The iconic case is probably that of Times Square, in the heart of
Manhattan, which has acted as a barometer of American attitudes
towards urban decline and inner city living (Reichl, 1999; Sagalyn, 2001).
Thus, central business districts are crucial attributes in the governance
of territory, and agglomeration, creativity and control are key themes in
understanding the significance of centrality in the contemporary
metropolis.

However, commentators are now increasingly thinking through models
of growth that reflect a sharply changing set of metropolitan dynamics.
Journalistic forays into these new landscapes revealed some fascinating
stories. Garreau's popular *Edge City: Life on the New Frontier* (1991),
collected a series of anecdotal accounts of social aspiration, fear of crime,
complex family commuting patterns and dislike for 'tax and spend' local

government. Joel Kotkin's *The New Geography* (2001) told a similar story a decade later, in the aftermath of the dot.com boom that did so much to reshape workplace and living patterns for high-tech employees. Urban theorists have not always been so interested in these landscapes, although Soja's (2000) *Postmetropolis* attempts to pull together patterns of deconcentration and recentralisation. This disrupts the conventional image of the city: as Soja (2000: 242) continues, 'the densest urban cores in places like New York are becoming much less dense, while the low-rise almost suburban-looking cores in places like Los Angeles are reaching urban densities equal to Manhattan'. These fragmented, decentred cities pose numerous problems for analysts. One major issue is that of definition: the following terms all connote differing versions and visions of post-metropolitan life: suburbia; cyburbia; edge city; autopia; non-place. Each implies a shift away from classical or even modern conceptions of the city, symbolised by the cathedral and the city square.

These metropolitian landscapes reflect the importance of speed to the constitution of city life (Hubbard and Lilley, 2004). The arrival of high-speed inner-city motorways and freeways has been a dominant feature of post-war urbanism, shifting residents' perceptions of the city. This often resulted in the displacement of long-established – and usually working class – communities, as in Berman's (1982) poignant recollection of the destruction of his childhood beneath the path of the Cross-Bronx Expressway in 1950s New York. Thus, the rhythmic nature of urban **public space** is important, not least in terms of how the needs of various users drawn to central places are resolved, planned for and regulated. The reclamation of human-scale streetscapes as a focus of collective memory in the urban arena has been an important trend (Hebbert, 2005). However, for the vast majority of cities, be it Beijing, Buenos Aires or Glasgow, the historic core remains a 'niche' within a broader city-region economy, with its specialised urban functions such as government offices, department stores, opera houses etc.

This has had an epistemological dimension too, challenging the often taken for granted notion of the 'city' as a coherent whole. For Amin and Thrift (2002: 8):

21

> The city's boundaries have become far too permeable and stretched, both geographically and socially, for it to be theorized as a whole. The city has no completeness, no centre, no fixed parts. Instead, it is an amalgam of often disjointed processes and social heterogeneity, a place of near and far connections, a concatenation of rhythms; always edging in new directions.

This 'anything goes' approach is exciting, but threatens to rattle itself to pieces if taken too far.

## The return to the centre: revalorising the 'zone-in-transition'

As described in the entry on **global cities,** major CBD office districts have become paramount in servicing the global economy. For Saskia Sassen:

> At the global level, a key dynamic explaining the place of major cities in the world economy is that they concentrate the infrastructure and the servicing that produced a capability for global control. The latter is essential if geographic dispersal of economic activity – whether factories, offices, or financial markets – is to take place under continued concentration of ownership and profit appropriation. (Sassen, 1995: 63)

Such concentrations have placed huge pressures on the existing urban fabric. This is particularly marked in cities with architecturally or physically distinguished cores. It is impossible to generalise about this. Cities as diverse as London, Barcelona and Rome have each retained vibrant central cities for a number of reasons (e.g., Herzog, 2006).

However, a key theme of locational theory was that of the 'zone-in-transition', a term coined by Ernest Burgess in his concentric ring model (see **diagram**). This term

> applies to that part of the central city which is contiguous with the CBD, is characterised by ageing structures and derives many of its features from the fact that it has served as a buffer zone between the CBD and the more stable residential districts of the city. (Herbert, 1972: 105)

Once dismissed as zones of small industry, poor quality environments and inferior accessibility, such areas have become highly sought after or 'revalorised' in the post-industrial city. A fundamental aspect of the revival of downtowns was their re-use as a *residential* neighbourhood, as social groups of varying degrees of affluence populated central cities. A key part of this rediscovery of downtown is the growing demand for inner-city residences, charted by theorists of gentrification (e.g., Ley, 1996; Smith, 1996). Allen (2007), reporting on research carried out in Manchester, England, identifies three principal groups that have (re)colonised these areas for different reasons. First, there are the

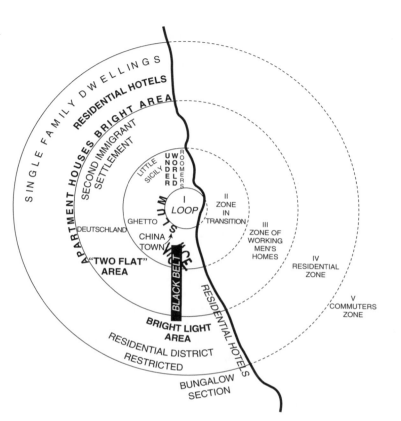

**Figure 1.1.2** Burgess' Model of Centrality

23

'counter-cultural' users, often associated with the city's nightlife. On the one hand, there is a 'creative class' of fashion designers, artists and architects; on the other, there is the gay and lesbian community, who have often carved out 'gay villages' in neglected corners of the city. Second, there are the 'successful agers', those who had raised families in suburbs and who had moved back to the downtown to enjoy its cultural facilities of restaurants and the arts. Third, there are 'city-centre tourists', those with a short-term aim of living in cities to enjoy the experience, but with a longer goal of moving to the suburbs. Each of these groups has different levels of cultural skill or capital, social networks and hobbies, and housing histories. But what united them was an interest in the city as a playground, a resource to be enjoyed. As Slater (2006: 738) argues, this has been mirrored in a shift in the focus of gentrification research: 'The perception [of gentrification] is no longer about rent increases, landlord harassment and working-class displacement, but rather street-level spectacles, trendy bars and cafes, iPods, social diversity and funky clothing outlets'.

The economic recovery of central business districts has led to some critics arguing that city centres are now dominated by suburban 'family' values. The density of city life and the vibrant socialities it affords retains a powerful hold over planners and policy-makers. Jane Jacobs' 1961 classic *The Death and Life of Great American Cities*, with its famous Manhattan-based 'sidewalk ballet' of dense sociality and pedestrian based 'street life', has influenced generations of well-meaning place-makers. It is ironic, then, that critics are now debating *The Suburbanization of New York* (2007), a collection of polemical essays about the future of what the editors claim to be 'the quintessential city: culturally, ethnically, and economically mixed, exciting and chaotic, elusive and spontaneous, sophisticated and endlessly creative' (Hammett and Hammett, 2007: 19). This city is seen to be under attack, a battery of writers asserted, by a combination of gentrification, in-town shopping malls, franchised national chain retail, festival marketplaces and Starbucks (M.D. Smith, 1996). Urban theory is notorious for being built upon a few unique cases and it is prudent to be cautious about the replicability of New York's experience. But many of the ideas raised above – of elite power in the centre, of the revitalisation, rejuvenation, renaissance of cities, of demographic 'invasions' by different social groups – can be traced out in slightly different forms in cities around the world during the 1980s and 1990s.

As noted, the revival of downtowns as a residential environment was driven by 'counter-cultural' or 'bohemian' social groups (Zukin, 1995). In cities such as New York, artists had gradually been rediscovering the warehouses and lofts of the industrial inner-city, drawn by cheap rents and raw, flexible industrial-sized floorplates. These partially remade city centres became more attractive to white collar workers, and – along with interventionist policing – downtowns became gentrified. This was given an academic rationale with the publication of Richard Florida's (2002) book *The Rise of the Creative Class* which became a near-instant hit with policy-makers internationally, welcoming of the message that successful regional economies had large numbers of bohemians, or 'artistically creative people' (p. 333), such as artists, designers or musicians. Increasingly, cities around the world have bought into the heavily marketed creative cities policy formula,

24

consciously branding themselves as buzzing, creative metropoles (Rantisi and Leslie, 2006). Reinvestment in the built environment is a key issue here. Conservation, anti-congestion measures and a reaction to modernist comprehensive redevelopment have reinforced and preserved the desirability and quality of life in old city centres (While, 2006). This has met with some criticism, however. On the one hand, the speed of transfer of this policy message has been attacked for its unrealistic assumptions (Peck, 2005). Furthermore, while some policymakers have seen night-time economy sectors as being 'a postmodern panacea of creativity', Hollands and Chatterton (2003) argue that this is a sector dominated by corporate entertainment companies that force out smaller operators: 'City centre nightlife', they suggest, 'is far from a level playing field and left to the market, many smaller-scale, locally-based nightlife producers are closed down, pushed to the margins, or simply bought out' (p. 380).

## Is centrality still important in cities?

To conclude, the city centre, downtown or urban core has a very strong presence within popular urban imaginaries. In one sense, developments in technology and the continuing desire to escape the congestion of central cities would still seem to render the need for centrality as illogical. Graham and Marvin (2001: 116–7) argue that in an informational age

25

> networked urban cores ... tend ... to be physically or technologically obsolescent, requiring significant and costly retrofitting. To the municipal jurisdictions in old urban cores, securing the financial and technological expertise to update infrastructure, whilst also facing the fiscal and social crises surrounding deindustrialisation and social polarisation, poses enormous challenges.

Yet the very problems of downtowns – of their age and apparent lack of suitability for contemporary living – are also their greatest advantages.

This is a simplified story, applicable primarily to US downtowns. Yet there has been a renewed interest in city centres in recent years in cities worldwide. Even the 'shock' cities of China are now increasingly interested in historic preservation and the adaptive re-use of existing buildings. A number of important processes are involved here: a complex interplay of restated civic identity with the privatisation of city services; a re-found interest in density and tall buildings; the space conflicts of pedestrianism and automobility; the return of residential living in downtowns and city centres; and a strong policy turn towards embracing 'playfulness' and creativity.

## KEY POINTS

- Central business districts have been an important part of traditional approaches to urban geography, as seen in models of urban structure and theories of location.
- There are subtle differences within such central areas, between high-rise office clusters, 'zones of transition' between service and manufacturing neighbourhoods, retail streets and apartment blocks.
- The 1980s witnessed the beginnings of a rediscovery of city centre living, characterised by apartment living and new social groups adding to a so-called 'urban renaissance'. However, critics have pointed to processes of displacement and gentrification, and the power of property developers and corporate leisure groups in shaping urban politics downtown.

## FURTHER READING

There are a lot of good social histories of downtown and city centre living. Along with the work discussed above, Paul Groth's (1994) book, *Living Downtown: The History of Residential Hotels in the United States* gives a fascinating insight to the culture of the rooming house that once predominated many American cities. See also Cocks' (2001) book, *Doing the Town: The Rise of Urban Tourism in the United States, 1850–1915,* which shows the longevity of popular desire to visit central cities for leisure. For an interesting perspective on the changing nature of office space in CBDs, see O'Neill and McGuirk's (2003) paper, 'Reconfiguring the CBD: work and discourses of design in Sydney's office space',

*DMcN*

# 1.2 MOBILITY

Cities buzz with movement. The flow of traffic, the back and forth of commuter trains, the screech of trams, the pulsing of millions of phone calls through copper and fibre optic cables, the step, step, step, of thousands upon thousands of people going about their business, the sway of bodies moving to a popular song in a late night bar or club. And this is to say nothing of the flow of food, drink, water and other provisions that provide cities with the basics of life. Nor of the counter movement of solid and liquid wastes that the modern cities produce each day. In fact, once you start thinking about cities as entities that are constantly in movement, the list of things that this movement – this mobility – includes is near to unending.

## Quantitative geography and transport geography

A central problem facing contemporary urban geography – and indeed the social sciences more generally – is how this buzz of movement should be accounted for. Of course, in all sorts of ways urban geography has been concerned with mobility from its conception. In the 1920s and 1930s the leading figures of the Chicago School of urban sociologists argued that a key element defining the emerging American industrial metropolis was its fluidity. Drawing inspiration from the German social philosopher Georg Simmel, the Chicago School emphasised how the life of the modern metropolis was defined in all sorts of ways by the circulation of people and information within *and* beyond the metropolis. Louis Wirth analysed the ways in which the urban dweller was defined through their mobility, adopting a range of different roles as they passed through the different parts of the city. Robert Park (see **community**) studied the ways the newspaper channelled the flow of information through the metropolis. In detailed case studies of the hobo, immigrant communities and itinerant communities of association like the taxi dance halls, the Chicago School spent a great deal of effort attempting to understand marginal groups whose character was defined to a great extent by their fluidity. Indeed, Ernest Burgess (in Sheller and Urry, 2000: 740) argued that mobility 'was perhaps the best index of the state of metabolism of the city. Mobility may be thought of in more than a fanciful sense, as the "pulse of the community"'.

A similar concern with movement can be found in the work of many of the pioneers of the quantitative revolution that swept through human geography – and with it urban geography – in the late 1950s and 1960s (see Ullman, 1957; Webber, 1964; Berry, 1973). One of the key concerns of quantitative geography was how distance and patterns of connection – road and rail networks, telecommunication systems and so on – influenced the amount of movement that occurred between different places. In particular, influential strands of quantitative geography focused upon how the constraints imposed by transportation – both monetary (it requires resources to move things from one place to another) and temporal (moving across space takes time) – structured the economic dynamics of economic development. Quantitative geographers developed a range of analytical models from simple locational analysis and gravity models, to linear programming and factor and network analysis to explore these spatio-temporal dynamics. And, they demonstrated that taking into account both the fact that economic activity had a geographic location, and that this activity is dependent upon the complex spatial-temporal coordination of labour, materials and energy, profoundly altered social science's understanding of how economies function.

28

Given the importance of transport to much of this quantitative work it is not surprising that in the 1960s and 1970s transport geography, or transportation geography, emerged into one of the central sub-disciplines within human geography (see Eliot Hurst, 1974; Hoyle and Knowles, 1992). And, as transport geography grew in significance within the discipline, the methods and techniques that it had developed simultaneously became influential in urban planning, urban design, and other areas of urban management.

## The mobility paradigm

Nonetheless, for all the success of quantitative transport geography, conceptually it offers a quite narrow interpretation of what the movement that animates cities is about. As Jean-Paul Rodrigue (Rodrigue et al., 2006: 5) and his colleagues write in *The Geography of Transportation Systems*, transport geography can be defined as being:

> concerned about movements of freight, people and information. It seeks to link spatial constraints and attributes with the origin, the destination, the extent, the nature and the purpose of movement.

This implicitly leaves out a great deal of movement that takes place within cities. It makes no mention of the circulation of water and waste upon which cities depend (see **infrastructure**). It does not take any account of the movement of non-human habitants of cities, plants, wildlife, pets, livestock, etc. (see **nature**) that are such an integral part of what cities are. Nor does it allow much space to think about all the movement that takes place in cities not oriented towards travelling from one destination to another; the pleasures of street cruising and hanging out in cars, Sunday strolls, dog walking or jogging, to name just a few examples. Put another way, transport geography is simply about getting from A to B, it is not about the journey (the movement) itself; it is oriented towards the narrowly instrumental. As such, the purely quantitative accounts favoured by transport geography tend to freeze the movement they are describing, reducing it to an abstract vector between two points.

That transport geography should frame movement in such a way is not inherent to the nature of the sub-discipline. As David Keeling (2007: 218) notes:

> Transportation is quintessentially geographic, so it seems surprising that anxiety still exists about whether transport geographers have strayed from the core theories and methodologies of geography. Have transport geographers lost touch with the core principles of their discipline in order to make their research more relevant to others? Are they stuck in the narrow confines of network structures and flows, unable to explicate the multiple ways that transportation shapes human activity across the globe?

29

Of course, it should be stressed that within the remit of transport geography the ways quantitative accounts flatten the experience of movement is not necessarily a problem. If the aim of transport geography is simply to describe movement within a defined transportation system and consider ways the efficiency (in terms of overall cost, average journey time, etc.) of this system might be improved, then focusing on the instrumental dimensions of transport is more than adequate. But, if our aim is to account for the buzz of movement mentioned in the introduction then there is a need to go beyond the boundaries of transport geography and the tools of quantitative analysis.

This is what the mobility research paradigm (Sheller and Urry, 2006) that has emerged in the past decade out of the work of a diverse range of social scientists interested in mobility and movement of all kinds seeks to do. Recognising – as transport geography did, as too in their limited way the Chicago School did – the centrality of mobility to the construction of contemporary industrial and post-industrial societies, a diverse collective of human geographers, sociologists and other social scientists have been attempting to undertake a thorough redefinition of what social science research should involve and what it should focus on. Extending, and radicalising, the insights, of quantitative geography that patterns of connection and movement define (or perhaps more accurately, *construct*) our social worlds in all sorts of fundamental ways, mobility theorists argue that social theory should start from the premise that the world is defined by movement and fluidity not by stasis and structure (see Thrift, 1993; Urry, 2000; Amin and Thrift, 2002; Sheller and Urry, 2004; Sheller and Urry, 2006; Cresswell, 2006). In the words of Mimi Sheller and John Urry (2006: 208),

> the social sciences have ... failed to examine how the spatialities of social life presuppose (and frequently involve conflict over) both the actual and the imagined movement of people from place to place, person to person, event to event. Travel [and movement] has been seen as a black box, a neutral set of technologies and processes predominately permitting forms of economic, social, and political life that are seen as explicable in terms of other more powerful processes. ... [A]ccounting for mobilities in the fullest sense challenges social science to change both the objects of its inquiries and methods of its research.

## Automobility and other mobilities

That sounds very abstract. And in many ways it is. Nonetheless, the sense of what Sheller and Urry (2006) are arguing for can be illustrated through the example of the automobile. One of the definitive objects of contemporary industrial society, the automobile is a mode of personal transportation, a signifier of status, an aesthetic statement, a symbol of individual freedom and the ultimate consumption good, as well as being a ubiquitous presence in contemporary cities (Sachs, 1984; Miller, 2001; Wollen and Kerr, 2002). Indeed, the philosopher Roland Barthes (1972: 88) went so far as to claim that:

> cars today are almost the exact equivalent of the great Gothic cathedrals ... the supreme creation of an era, conceived with passion by unknown artists, and consumed in image if not in usage by a whole population which appropriates them as a purely magical object.

**Figure 1.2** 'Automobility'

But the automobile is also rather more than just a technological object, or a symbol of a certain kind of society. In a very real sense the automobile defines the patterns of social organisation and interaction that characterise contemporary urban life: the automobile has become in a quite concrete sense a whole way of life.

And, yet, the social sciences, from urban geography, through to sociology and even urban studies, has failed, as Sheller and Urry (2006: 209) write:

> to consider the overwhelming impact of the automobile in transforming the time-space 'scapes' of the modern urban/suburban dweller. Industrial sociology, consumption studies, transportation studies and urban analyses have each been largely static, failing to consider how the car reconfigures urban life, with novel ways of dwelling, travelling, and socialising in and through an automobilised time-space.

What is more, the more one thinks about the place of the automobile and the mobility it affords, the more all embracing its presence appears, the more it appears to 'unfold a specific pattern of domination' (Heidegger, in Sheller and Urry, 2000: 737):

> Automobility impacts not only on local public spaces and opportunities for coming together, but also on the formation of gendered subjectivities, familial and social networks, spatially segregated urban neighbourhoods, national images and aspirations to modernity, and global relations ranging from transnational migration to terrorism and oil wars.

To take the *automobilisation* of urban life seriously, therefore, requires the recognition that the success of the automobile is based on the intricate network of relationships that the automobile draws around it.

The most obvious dimension of drawing together is the infrastructural ecology that the automobile is dependent upon: wide smooth roads, gas stations, automobile repair shops, car dealerships, parking lots and buildings, traffic control systems, auto recovery companies, to name just a few obvious examples. There is another, equally intricate, socio-legal ecology of insurance companies, government statutes, local by-laws, accident claims companies, traffic police, parking wardens, that has grown up to regulate, order and, where necessary, discipline the masses of cars (or to be more accurate, car-human hybrids) that populate contemporary cities (see Beckman, 2001; Jain, 2004; Latham and McCormack, 2004; Merriman, 2007; Dodge and Kitchin, 2007). On a more affective level, the automobile has become profoundly implicated in people's emotional lives in all sorts of ways. Jack Katz (1999) has written about the distinctive forms of affective attachment that drivers in Los Angeles have developed for their cars. To disrespect someone's road space is to disrespect in some very real sense their personal space. While, to take another example, Robyn Dowling has described how for many suburban women, car ownership has not only become pivotal to the way they manage the daily working and childcare routines, but the car and its interior space have come to embody the sense of care and love mothers have for their children. They come to embody a widely recognised element of 'good mothering' (Dowling, 2000: 352; see also Law, 1999).

Of course, the automobile is by no means the sole armature of mobility in the contemporary world. Peter Adey (2007), Tim Cresswell (2006), and Martin Dodge and Rob Kitchen (2004) have described the ways that air travel has become an increasingly pervasive element of urban life, generating a novel range of new spatial-temporalities.

Indeed, in all sorts of ways airports with their hotels, offices, mass-transportation systems and shopping malls, are coming to resemble mini-cities. And other writers have shown how the mobility inherent in practices such as mass tourism (Sheller and Urry, 2004), transmigration (Conradson and Latham, 2005a, 2007), the use of mobile telephony (Katz and Aakhus, 2002), or the internet (Wellman, 1998; Wellman and Haythornthwaite, 2002), structure contemporary urban life in all sorts of interesting ways. And all this is to say nothing of the phenomenal movement of goods, commodities and information that globalised cities both organise and are dependent upon (see Appadurai, 1996; Castells, 1997; Amin and Thrift, 2002). In fact, the more one looks, the more one finds the potential to understand all urban phenomenon as in some sense mobile.

# Central propositions of the mobility paradigm

The central point of the mobility paradigm is that it is both possible and productive to interpret cities as organised through multiple forms of movement, rhythms and speeds. And while things like automobiles, motorways, airplanes and airports are the most obvious dimensions of the 'mobility turn', the real challenge of the mobility paradigm is its demand to 'open up all sites, and places, and materialities to the mobilities that are already always coursing through them' (Sheller and Urry, 2006: 209). The question, then, is what exactly does that mean for how urban geography should research cities? One way to answer this question would be simply to continue to list all the ways that cities are defined through mobility. Another way to address this is to draw up a list of general conceptual propositions that define the new mobility paradigm as a way of thinking about the social world. So, at the risk of over-simplification, we can summarise the mobility paradigm into eight key propositions:

1  The world is defined by motion and fluidity, not primarily by stasis.
2  The world is made up of a heterogeneous multitude of time-spaces.
3  The social sciences need to become post-humanist. The social is made up of non-human *and* human actants.
4  Social theory/social science needs to move beyond the nation state. That is to say, social theory should not be based on the assumption that the nation state is the natural home of society.

5   To understand society, social scientists need to focus on social practice.
6   To understand society, social scientists need to focus not just on rationality, but also on the affective dimensions of life.
7   Society needs to be understood through and through as socio-technical. The world of technology is not something outside of 'the social'. It is implicated fundamentally in the social world's unfolding.
8   The principle task of social theory as a kind of metaphor making. The kinds of metaphors used to describe reality come, in all sorts of ways, to define this reality. Metaphors also delimit how we understand problems. So in place of metaphors like structure, agency, etc., mobility research is organised around metaphors of flow, fluidity, network, scapes and complexity.

Exactly where the mobility paradigm's conceptual propositions will take urban research is unclear. The turn towards mobility within the social sciences in general – and urban geography in particular – is still relatively recent, and the mobility paradigm remains a diffuse and rapidly evolving intellectual movement. Nonetheless, in prompting urban geographers to re-consider such fundamental notions as **community** (what does it mean when community is mediated through the mobilities of the automobile, or telephone, or internet?), or conceptions of public-ness and **public space** (has the automobile's domination of the urban street killed public space, or transformed it creating a new public of automobility?), and in bringing in all sorts of neglected and overlooked aspects of urban life to researchers' attention (the centrality of all sorts of **infrastructures** to the smooth functioning of cities, the diverse socialities associated with movement, the ways mobility shapes peoples' mental map of a city), the mobility paradigm has helped to re-energise and reanimate, urban geography.

## KEY POINTS

- Movement is one of the defining elements of urban life.
- Quantitative urban geography and transport geography provided one of the earliest and most intellectually adventurous approaches to studying cities and mobility.
- Quantitative transport geography, for all its strengths, provides only a relatively narrow, instrumentally oriented, account of the movement that structures city life.

- The mobility paradigm is an interdisciplinary collective of researchers attempting to reconfigure the social sciences (including human geography) through placing movement and mobility at the centre of the social science's definition of society.
- Automobility offers an exemplary case study of what the mobility paradigm involves. It shows that the automobile is more than just a technological object – it is in a very real sense 'a whole way of life'.
- Rather than being defined by specific themes or research objects, the mobility paradigm is better thought of as a connected series of presuppositions about how to approach social scientific research.

### FURTHER READING

Tim Cresswell's (2006) *On the Move: Mobility in the Western World* is an engaging and thought-provoking study of the place of mobility in western society by one of human geography's most prominent theorists of mobility. John Urry's (2000) *Sociology Beyond Society: Mobilities for the 21st Century* offers a programmatic statement of what a 'mobile' social science involves. Despite what the title might suggest, the book is as much about a human geography of mobility as it is a about a sociology of mobility. Mike Featherstone, Nigel Thrift and John Urry's (2005) edited collection *Automobilities* is a compelling collection of essays exploring the role of the automobile in contemporary society.

35

*AL*

# 1.3 GLOBAL CITIES

The growth of globalisation as a field of research has been felt across many of the social sciences, from economics to anthropology to urban planning and – given its intrinsic spatiality – it has been a major preoccupation of human geographers. Yet the nature of this beast has been difficult to pin down, particularly when it is applied to the study of cities. Generally speaking, there are two schools of thought on the nature of global cities. On the one hand, there is a model based upon scale, network and hierarchy. This has been seen as based upon a static, fixed notion of the world, where cities can be identified as a single entity, then categorised, compared, contrasted and debated. The other school of thought sees globalisation as a process, disputes an easy identification of cities as objects and emphasises flows and movements between actors operating in the spatial formations that we know as cities. This division has been debated vigorously within geography and other disciplines, and the discussion that follows is an inevitably brief attempt to chart out some of its contours.

# From world city to global city

Although theorists such as Peter Hall (1966) and Jean Gottmann (1957) had provided a significant overview of the growing power of major cities, the emergence of world and global cities as a self-knowing concept is usually dated to the research agenda set out by John Friedmann and Goetz Wolff (1982), in an article published in the *International Journal of Urban and Regional Research,* which remains one of the key arenas for debates around the topic. Their argument brought together a number of 'macro' theories which aimed to explain globalisation, such as political economy and world systems approaches, and provided a succinct overview of the structural forces that were, they argued, placing cities – understood as expanded metropolitan city-regions – at the centre of economic dynamism, rather than the apparently declining centrally-coordinated economies of nation-states. Their agenda centred around the importance of cities as 'command and control' centres, clustering together the decision-makers of the transnational corporations that were exerting an ever-increasing grip on the world economy. Their argument set out a series of key restructuring processes – economic, social, physical, political – which are often difficult to disentangle, yet which taken together suggested a near-total transformation of how cities are theorised. Interestingly, the case study that inspired Friedmann and Wolff was Los Angeles, a spatial formation characterised by key post-industrial economic sectors such as military technology and film production. LA's predominance in these new theoretical debates – developed by others such as Ed Soja, Jennifer Wolch, Allen Scott, and Mike Davis – has remained controversial, critics pointing to the specificities rather than its archetypical characteristics. This point is revisited below.

Friedmann followed this agenda with 'The world city hypothesis' (1986), and by the late 1980s regional development theorists, urban planners, sociologists and geographers were deepening its theoretical complexity and empirical resonance. However, perhaps the key event in shifting from the 'world' city was the sociologist Saskia Sassen's *The Global City: New York, London, Tokyo* (1991), which, as the title implies, shifted concern to the three key cities as the apex of the increasingly financialised world economy of the 1980s. The novel aspect of Sassen's argument was that the mere possession of corporate headquarters was not enough for cities to dominate in the global economy. Rather, it was the cluster of interacting producer services firms that gave cities their ability to influence global flows. In addition, Sassen suggested that 'the difference between the classic concept of the world city and the global city model is one of level of generality and historical specificity. The world city concept has a certain kind of timelessness attached to it where the global city model marks a specific socio-spatial historical phase' (2001: 349). Thus, while world cities could be identified from at least the rise of world imperialism, Paris, London and Lisbon being major command and control centres in earlier centuries, Sassen was charting the specific locus of her chosen three global cities in terms of their power in shaping or coordinating the world economy of the 1980s and 1990s. The concept of global city was thus seized upon by policy-makers, business and academia alike, as it arrived at a moment when globalisation had firmly arrived on the agenda as a vital area of study.

37

The global cities agenda had reached such intensity by the mid-2000s that a major international publisher, Routledge, had commissioned a 'Global Cities Reader' running to 50 extracts from a wide range of perspectives (Brenner and Keil, 2006). Scholarly debate on the area had expanded into fields of representation, cultural identity, and the nature of the concept as a geographical problematic, as well as quantitative empirical testing, which reflected its purchase among scholars, not only those working in the field of the urban. In the remainder of this discussion, several of these areas are briefly addressed: the development of the global cities agenda on the basis of empirical mapping; the physical transformation of cities associated with the concept; critiques of the concept on the basis of its geographical prejudice; the emergence of 'global city' as a public discourse used by agents such as politicians and economic planners; and the widening of the concept with its explicit application to fields such as disease, religion and fashion that are not primarily associated with economic organisation. Elsewhere in this book we discuss how the cognate fields of **transnationalism** and **mobilities** have developed and deepened understandings of distanciated social identities and material and embodied travel.

# Cities and global economies

In an attempt to deepen understanding of the specific, measurable activities that link cities together in a global economy, a significant research network emerged around the Globalisation and World Cities (GaWC) cluster at Loughborough University in the late 1990s in the UK. As Beaverstock, Smith and Taylor (2000) have argued, a new 'metageography' of inter-city flow can now be identified and mapped. This entailed a categorisation of cities based on their relative connectivity, measured in material terms by data on internet connections, airline passenger traffic and office location patterns, particularly in advanced producer services such as law, accountancy, advertising and finance. As they argue:

> World cities are not eliminating the power of states, they are part of a global restructuring which is 'rescaling' power relations, in which states will change and adapt ... The 'renegotiations' going on between London's world role and the nation's economy, between New York's world role and the U.S. economy, and with all world cities and their encompassing territorial 'home' economies, are part of a broader change affecting the balance between networks and territories in the global space-economy. (Beaverstock et al., 2000: 132)

Thus, a sense emerges here that such cities are no longer articulated towards leading their national economies, that they are agents in themselves coordinating flows that will possibly bypass and probably even 'leak' from, the national economy in a number of ways.

This shift has been accompanied by a retheorisation of how cities are imagined. Amin (2002), seeking to make sense of the distanciation of economic transactions (in other words, the stretching of social interaction over many miles and continents), poses two viewpoints as to how cities can be understood. First, they can be seen as 'a string of place-based economies' (p. 392), a set of bounded territories which can be literally measured, for example in square footage. In contrast, Amin sees cities 'as a site of network practices' by which he means that the city is a 'nexus of economic practices that does not return the urban as a place of localised transactions'

(p. 393). In this view, 'the city thus conceptualised is no longer a bounded, but spatially stretched economic sphere' (p. 393). Amin's argument is thus based upon a finely grained distinction as to how we see cities: not, as he puts it, 'islands of economic competitiveness or knowledge formation', but rather as 'circulatory sites', restless and unstable (p. 395). Thus we cannot *assume* that cities are cohesive entities. As with Sassen's intimation about the importance of producer service firms (such as PricewaterhouseCoopers, for example), it may well be the transnational firm and its organisational linkages that define a city's cohesion.

Whichever side one takes on this debate, it should be stressed that despite the abstract diagrams or tables that identify these global cities, there are specific clusters or sites where interactions (often face-to-face and embodied) take place. Here we can see the significance of fixity and locational specificity as opposed to pure flow. This has often been related to the nature of office markets in such cities, characterised by the popularity of high-rise buildings and the tendency for advanced producer services – such as law, accountancy, insurance, architecture and advertising – to cluster around major commercial sites, thus intensifying inter-firm interactions. This has reached such an intensity that the design of both interiors and exteriors (O'Neill and McGuirk, 2003; McNeill, 2007) of office buildings is given great importance, with elite architectural firms competing fiercely for such commissions as the Swiss Re tower in London. Furthermore, the social practices that these buildings house have been the subject of some fascinating research, such as on gendered work practices in investment banks (McDowell, 1997), elite social networks such as expatriate clubs (Beaverstock, 2002), and the spatialisation of money and finance (Leyshon and Thrift, 1997).

39

## Global city discourses

There is another way of considering this, which is to see global cities as self-generating discourses. Influenced by the broad sweep of post-structuralism through geography and sociology, academics began to take greater care in how apparently neutral concepts influence real social action. In other words, once mayors, policy consultants, firm strategists, academics, journalists and others involved in knowledge circuitry agree upon the idea of a 'global city', then the term acquires a practical significance. Policymakers identify 'globally competitive' sectors; land use planning decisions are driven by a 'global' agenda; the fear of foreign competitors can be raised as a means of justifying controversial policy options. Ken Livingstone, Mayor of London, (2000–2008) constantly emphasised the need for skyscraper office buildings because of the imagined challenge from Paris and Frankfurt, for example. As Machimura (1998) has described, Tokyo's city government rescripted the city (and its land use planning) to allow it to host and connect with global flows, using a homogenised language (perhaps emerging from a largely uncharted discourse provided by global consultancy firms) (see also McNeill et al., 2005 on Sydney; Paul, 2004 on Montreal).

A further issue is that while these small areas of capital accumulation – we could call them footprints – may have huge impacts on surrounding housing and labour markets, it is nonetheless true that the excessive focus on financial clusters may divert attention from mundane, but in aggregate terms significant, economic zones such as small businesses, distribution parks, informal economies of unpaid or semi-legal work, or even criminal economies. This is particularly important when considering the importance of underdeveloped cities in Africa, Asia and Latin America, which may plug into global economic flows in less spectacular ways (Sidaway and Power, 1995; Shatkin, 1998; Simone, 2004).

## Disease, religion, fashion, art

However, it should be noted that an over-concentration on financial services may obscure the many other relationships that bind cities, or sites within cities, together. To illustrate this, consider the following four areas: disease, religion, art and fashion. The global geographies of disease were illustrated with the sudden explosion of the SARS epidemic in 2003, a 'transmission chain' that linked rural China with urban Canada. As Harris Ali and Keil (2006) have shown 'the disease originated in rural southern China, then moving to major cities in China via live animal markets, eventually to find its way to the major global city of Hong Kong, which then served as an important interchange site for its global spread' (p. 500). The peculiar geography of SARS – with a concentration of clusters in Southeast Asia but with an outlier in Toronto – highlights the increasing trade and travel links between Chinese cities and the primate Canadian metropolis. As Harris Ali and Keil continue, while European port cities were always significant transmission points for disease in the pre-modern and modern periods, the contemporary situation has aided the spread of epidemics. This is for two reasons: first, 'the incubation period for many viruses and bacteria is much shorter than the travel time for transcontinental aircraft trips ... those affected are very likely to be asymptomatic during their trip and upon arrival in their destination' (p. 492); second, 'the routes through which pathogens can enter the populations of cities have multiplied' (p. 492) due to global city networks.

It is now well-established that religion is a great shaper of the world, that the 'world religions', particularly those of Islam and Christianity, have very significant effects on the social practices, cultural values and legal norms of nation-states. The Roman Catholic Church, for example, which is probably the world's largest institution in terms of practising members, is very reliant on the territorial power base and mytho-logisation of Rome as a direct lineage to Christ on earth. This has a huge impact on the everyday practices within the secular city, along with its urban politics (McNeill, 2003b). In the extended social and cultural networks found in most major cities, religious institutions have key functions in providing a focus for diaspora groups. The mythic, essentialised nature of sacred Rome mobilises hundreds of thousands of visitors and pilgrims; this aspect is in turn relational in that it ties Rome into networks of faith communities in all parts of the world; and it then requires the institutionalisation and regulation of these networks in a way that must balance sacred claims to space with the secular rights of Roman and Italian citizens. Similarly, the construction of religious buildings such as mosques – and local responses to these events – reveal deep-seated cultural identities and anxieties which both feed from and structure national imaginaries (Dunn, 2005).

41

A further example is that of fashion, and particularly that of high fashion. As with religion, certain key cities have historically dominated this movement, as Gilbert (2006: 20) describes:

> The growth and systematization of European imperialism was an important phase in the development of fashion's world cities ... London and Paris came to be understood as sites of both innovation and of fashion author-ity. This worked through the actual export of clothes and designs, but also through the symbolic projection of these cities as avatars of fashionable modernity.

In a similar way, London emerged as a major world centre for modern art in the 1990s. For While (2003),

> the business of buying and selling Western high art is dominated by New York, Paris and London, as well as a number of second-order international

> nodes such as Los Angeles, Tokyo, Zurich, Milan and Dusseldorf ... the key
> international cities have become the home of the most influential interna-
> tional dealers, auction houses, critics and galleries, and act as magnets for
> aspiring artists and dealers, who in turn further enrich the creative milieu of
> art schools, galleries and cultural quarters. (p. 253–4)

What these two sectors have in common is the transmission of cultural values, where the evolution of complex systems of art and fashion appreciation has become housed in the major institutions of certain world cities.

It should be said that each of these fora for cultural exchange are partially constituted by the material infrastructures that sustain them. Whether this be the human or animal carriers of disease, the rocky pilgrim routes of the Middle Ages or the charter flights that transport adherents to Mecca or Rome, the globalisation of architects and their designs (McNeill, 2008a), or the networks of art auctions and fashion magazines that sustain and transmit cultural artefacts, the material nature of this connectivity is very important (see **materiality**).

# 42 Positionality

Finally, a powerful critique of the global cities literature has emerged from geographers charting the global South. Writers such as Robinson (2002), Gandy (2005a), and Simone (2004) have argued that urban theory-building has ignored the specific conditions of African cities, or – worse still – have even exoticised their urban experience. For Robinson, the problem lies in the division between economic and urban geography as disciplines that focus on advanced capitalist economies, and development geography as focusing on less developed countries:

> In the same way, then, that global and world city approaches ascribe the
> characteristics of only parts of cities to the whole city through the process
> of categorization, mega-city and developmentalist approaches extend to
> the entire city the imagination of those parts which are lacking in all sorts
> of facilities and services. (Robinson, 2002: 540)

The effect of this, for Robinson, is to pathologise such cities as being poor, and requiring external help, ignoring the potential to redistribute wealth (which is concentrated in certain zones within African cities) internally within cities. As Shatkin (1998) has argued, seeing certain cities and nation-states as being irrelevant to global cities research due to very low levels of foreign direct investment is to ignore the increasing integration of cities such as Phnom Penh or Nairobi within the world economy.

Theorising global cities is thus a major issue within contemporary social sciences. The quantitative analysis of major data-sets which show degrees of inter-connection and the magnitude of inter-city flows of anything from people to food have made a major contribution to our understanding of the globalised urban economy. However, the broader remit that an urban cultural perspective brings serves to deconstruct the categories used to make knowledge claims, challenge the over-emphasis on certain cities as being foundational in world city analysis and to bring non-economistic indicators into the theoretical armature of urban geographers.

## KEY POINTS

- The debate has often been focused around economic globalisation, potentially neglecting the importance of cities as importing and exporting artefacts in fields such as fashion, art, architecture and religion.
- Critics have suggested that the global city term focuses attention on key financial centres, to the neglect of underdeveloped, yet incredibly populous, urban areas around the world.
- The terminology used in the debate is fundamental. It is not clear that 'city' is an appropriate term with which to frame the debate, with many theorists preferring to speak of 'sites' of networked activity.

43

## FURTHER READING

There are two edited collections that cover the field in considerable depth: Knox and Taylor's *World Cities in a World-System* (1995) and Allen Scott's *Global City-Regions* (2002). Jenny Robinson's (2006) *Ordinary Cities: Between Modernity and Development* provides a much-needed bridge between the global cities literature and theoretical perspectives drawn from development geography.

*DMcN*

# 1.4 TRANSNATIONAL URBANISM

Transnational urbanism is a term originally coined by the American anthropologist Michael Peter Smith (2001, 2002, 2005a). It is an attempt to think through the ways in which cities are evermore defined by all sorts of connections to faraway places. More specifically, it is an attempt to think systematically about the ways such long distance – often trans-border – connections are increasingly organised through people leading lives that are lived in 'two places at once', lived both "here *and* there" (Smith, 2005b: 82, *emphasis added*). This fact of lives lived in two places at once is significant for at least two reasons. First, social relationships that are routinely stretched over long distances place into question in all sorts of ways how social researchers should understand the 'there-ness' of social interaction. Second, the spatially distributed agency implicit in such geographically stretched networks of relationship raises questions about the nature and role of cities as nodes of concentrated interaction. If a great deal of the interaction taking place within cities is in fact organised through relationships from elsewhere, what does this do to established notions of cities as privileged central places? Does it mean that the nodality of cities no longer matters? Or that it matters less than it has in the past? And if it matters less, what does this do to established ways of under-standing cities? Does it mean that they are redundant? Or just that they need a little rethinking? The concept of transnational urbanism is an attempt to place these issues of spatial distanciation at the centre of how cities should be understood. Rather than thinking about cities as princi-pally contained – and indeed easily located – sites, transnational urban-ism suggests that urban theory needs from the very start to recognise that cities are constituted through often extraordinarily complex time-space entanglements – entanglements in which the notion of what is 'near' and what is 'far' is often very much less than self evident.

## Transnationalism and the transnational

Although used as early as 1916 by the American essayist Randolph Bourne, the term transnationalism as currently understood has its origins in work in international relations and economics in the 1960s (Bourne, 1916; Keohane and Nye, 1971). Scholars in these fields used the term transnational to refer to institutions and forms of relationship that spanned national borders and in some way transcended the national. It was not until the 1990s, however, that the term transnational – and the related noun transnationalism – gained genuinely widespread currency. A range of scholars working in anthropology (Sutton, 1987; Glick Schiller et al., 1992; 1995; Smith, 1994; Hannerz, 1996), sociology (Levitt, 1994; Portes, 1996), human geography (Mitchell, 1997), literary studies (Tölölyan, 1991, 1996) and elsewhere began arguing that the concept of the transnational offered a productive way both for addressing the ways in which the world was becoming increasingly globally interconnected, while at the same time acknowledging that the nation state still remained an important part of this globalising world. As Smith (2001: 3) writes:

> ... Globalization discourses ... often explicitly assume the growing insignificance of national borders, boundaries and identities. In contrast, transnationalist discourse insists on the continuing significance of borders, state polities, and national identities even as these are often transgressed by transnational communication circuits and social practices.

**45**

In particular, much work on transnationalism came to focus on what Nina Glick Schiller, Linda Basch and Cristina Blanc-Szanton (1995: 48) called 'transmigrants'. Transmigrants are international migrants who – in contrast to earlier patterns of migration, 'live dual lives: speaking two languages, having homes in two countries, and making a living through continuous regular contact across national borders' (Portes et al., 1999: 217).

This pattern of transnational migration is a phenomenon that has historical precedents. Far from being a 'melting pot', for example, late nineteenth and early twentieth century American society was much more a mosaic of lingering national and ethnic affiliations. (This, in fact, was the topic of Bourne's 1916 essay, just as it was the focus of much of the Chicago School's research). Nonetheless, students of contemporary transnational migration argue that contemporary forms of migration are in important ways qualitatively different from earlier patterns. This is for three reasons. First, the kinds of connection possible in the contemporary 'transnational moment' (Tölölyan, 1991: 5) are profoundly different to those available to previous generations of migrants. Technologies like cable and satellite television, long distance telephony, jet aircraft – to say nothing of the internet – allow a quality and immediacy of interaction that simply was not possible with earlier technologies. Second, the intensity and duration of the level of connection is much greater than previously. It is not just that long distance communication and transportation has gotten faster. The cost of maintaining links between a migrant's old and new homes has fallen precipitously over recent decades. This allows even the poorest and most socially disadvantaged to be involved in circuits of transnational exchange that previously would have only been the preserve of elites. Third, contemporary migration is taking place at a time when many migrants not only come from countries with a well developed national identity but also with a well defined desire to maintain that sense of identification. Thus, many migrants *despite* migrating are prepared to actively defend their own national identity in the face of efforts by receiving country governments to assimilate them into the host culture (Glick Shiller et al., 1995; Portes et al., 1999).

# Transnational cities, transnational social morphologies

46

What does all this mean for how contemporary cities should be understood? Well, according to Smith (2001) and other scholars of transnationalism the emergence of complex networks of transmigration matters for how cities should be understood for at least three reasons:

1  It matters because cities are where the great majority of contemporary migrants live.
2  It matters because the presence of transmigrants profoundly shapes the dynamics of much contemporary urban life. They influence labour markets, politics, the nature of a city's international connections, to say nothing of its cultural and public life.
3  It matters because cities are the central 'sites for concentrating the social, physical, and human capital used to forge other types of [transnational] socio-economic and political projects across borders' (Smith, 2005a: 7).

Taken together, these three factors point to an emergent transnational urbanism, an urbanism that is constituted through the interaction of a complex – and interlinked – range of what Steven Vertovec (1999) calls 'transnational social morphologies'. The key dimensions of this transnational urbanism can be summarised under three general headings: transnational social fields; transnational economies; and transnational political formations.

## Transnational social fields

Perhaps the most striking dimension of the phenomenon on transmigration is the degree to which it has become possible for people to construct remarkably dense and complex kin and friendship networks across what for an outsider appear impossibly large distances. Thus, Glick Shiller et al., (1995: 54) describe the case of a family from the Caribbean island of St Vincent where:

> Two daughters, who could not find employment in St. Vincent … migrated to the U.S. as domestic workers to gain income to support family members in St. Vincent and contribute to building a cement block family home. Two brothers, who also could not find work locally, migrated to Trinidad as a skilled automobile mechanic and construction worker. The wife of one of the brothers later joined her husband's sisters in New York, where she too became a live-in domestic worker. The mother remained behind in St. Vincent to care for her son's two small children and oversee the construction of the family home.

47

This pattern of interlinked family movement is characteristic of many relatively impoverished 'transnational' or 'translocal' migrants. Transmigrants improvise family structures that stretch across often multiple national boundaries, balancing off as best as they can the different economic, social and legal possibilities presented by different places. So, while St Vincent does not offer sufficient economic opportunities for the economically active members of the family described above, it does offer a base from which the family's children can be cared for while their parents work outside of St Vincent. And while New York may only offer a precarious living for the women of this family, and little chance of permanent settlement, it does at least offer American wage levels and payment in American dollars.

But these transnational networks take in more than just families. As the sociologist Peggy Levitt (2001) has shown, they also incorporate whole neighbourhoods and in the case of the 'sending' country often whole towns and villages. And, it is not just poor people from less-developed countries implicated in the construction of these transnational social fields. At the other end of the social scale Leslie Sklair (2001) has documented the emergence of a 'transnational capitalist class', who are the managers and controllers of the contemporary global economic system. Despite their power, even this group cannot fully escape the costs of frequent long-distance movement. They too have to find ways of aligning the demands of their often highly mobile lifestyles with those of family life. Indeed, as Aihwa Ong (1999; see also Mitchell, 1995; Olds and Yeung, 1999) has shown in her research on Overseas Chinese entrepreneurs, the development and sustenance of often intricate transnational family networks have become a key dimension of many Overseas Chinese business ventures. In a similar vein, Johanna Waters (2005, 2006) has mapped the transnational strategies that professional upper-middle class Hong Kong Chinese families employ to gain maximum educational advantage for their children – sending their children to costly overseas English language universities in Canada, the UK and the US. Together such transnational strategies have generated a rich new typology of social forms – astronaut fathers and families, parachute kids, satellite children – as family heads try and manoeuvre their family resources across international borders to best advantage.

## 48  Transnational economies

To speak of a transnational capitalist class is to point to how transnational social fields are intimately intertwined with the economics of the global economy. Now in Smith's (1998, 2001) account of transnational urbanism the global economy per se has a limited role. As he points out, many of the international flows of capital that are reshaping supposedly iconic global cities like Los Angeles are in fact far from global. Rather, such flows tend to trace out a more restricted trans-regional structure. Downtown Los Angeles has been rebuilt over the past couple of decades with money from Japan, Hong Kong and the Middle East; not money from just anywhere. But this is not to say in a more general sense that the criss-crossing flows of international commerce have not profoundly reshaped certain areas of cities. In most large cities it is possible to trace out an infrastructure of elite transnational mobility afforded through things like 'hotels, airports, and similar institutions', which, as the anthropologist Ulf Hannerz puts it, 'are intensely involved in mobility and the encounters of various kinds of mobile people' (Hannerz, 1998: 239).

Spaces like airports and hotels *are* important in the emergence of a transnational urbanism (see **mobility**), but Smith (2005a: 9) is also right to ask researchers to look beyond 'the hypermobility of key sites' such as these in trying to understand transnational urbanism. So, for example, and moving away from the built form of the city, Phil Crang, Claire Dwyer and Peter Jackson (2003; Crang and Dwyer, 2002) have traced the transnational circuits through which ethnic entrepreneurs bring together, interpret and reinterpret products and materials from their different 'home' cultures to develop new styles of fashion, ways of eating and more. What is significant about these transnational entrepreneurial networks is how they sit between two or more places and how they reach out beyond the world of the transmigrant into 'mainstream' urban culture.

Of course, there are more prosaic dimensions to this transnational entrepreneurialism. There are the legions of transmigrant entrepreneurs who provide the everyday services and products that migrant communities rely upon, the everyday infrastructure that sustains and supports the transnational social fields discussed previously – the grocery and international phone stores, the internet cafes and travel agents, the money exchange services, the foreign language newspapers, to name just a few examples (Levitt, 1994; 2001; Smith, 2001; Friesen et al., 2005). Nor should we forget the profound transformation of existing urban neighbourhoods that the presence of large numbers of transmigrants brings with it. Writers like Mike Davis (2000), Michael Dear (2000; Dear et al., 1999), and Margaret Crawford (1999) have documented how the 'Latinization' of cities in states like California, Texas and Florida, has re-enlivened the public culture of cities as diverse as Los Angeles, San Diego, Miami and Houston, not least through the presence of large numbers of street vendors, yard markets, home based businesses like barbers, or second hand clothing stores, and other novel 'informal' economic strategies that migrants have brought with them.

Lastly, in focusing on transnational economies, we must not forget the importance of the labour that many transmigrants provide – this is after all the principal reason most migrants move in the first place. Migrant workers not only pay a central role in the labour markets of many so-called global cities like New York, London, Singapore and Paris, they are also increasingly a crucial part of the economies of less globally prominent cities. As we have already observed, mobile, highly educated professionals – computer programmers, engineers, lawyers, financial and management experts, the cadres of Sklair's (2001) transnational capital class – play a key role in knitting together global supply chains (see Beaverstock, 2005). Equally, a whole range of other occupations from bus drivers, cleaners, nannies and gardeners, to nurses and nurse aides, to waitresses, cooks, dishwashers and security guards, to taxi drivers and construction workers are frequently dominated by migrant labour. While much of the emphasis in the current literature focuses on low paid and often unskilled workers – which certainly does constitute a very significant proportion of this non-elite migrant labour force – the significance of skilled and often highly educated people like teachers, or doctors, should not be overlooked (Sassen, 1988; Waldinger, 1999; Erhenreich and Hochschild, 2003; May et al., 2007; Datta et al., 2007; Conradson and Latham, 2005a).

## Transnational political formations

The final dimension of the emergent transnational urbanism high-lighted by Smith is the political. One might think that transmigrants living between two different places would have little space for politics. If, in the famous words of the American congressman Tip O'Neill, 'All politics is local', one has to ask where is the 'local' in a life lived spanned between different places? It is precisely this question that makes the politics of transnationalism interesting – it is a politics that relocates the realities of political action, and places into question truisms such as O'Neill's statement.

It is possible to map out at least three modalities through which transnational political action is shaping contemporary cities. First, trans-migrants, despite migration, often remain linked closely concerned with politics in their country of origin. At the most basic this might simply mean migration creates a pool of ex-patriot voters. However, in many cases it also involves the creation of political constituencies that have subtly different aspirations, claims, or hopes, than those in the 'home-land'. Perhaps the iconic image of this kind of transnational politics is that of the political exile or refugee who campaigns for freedom and democratic rights in their homeland. Yet, as Smith (2001; Smith and Bakker, 2005) has shown, by no means all diasporic political movements are of such a politically liberal nature. Indeed, one of the most interesting dimensions of many contemporary transnational political organisations is how many are directed at 'modernising' and 'marketising' their homeland societies. In any case, what matters for the city in which trans-migrants have made their home, is that much of the political action concerned with the trans-migrants' original home actually takes place within, and circulates around, the migrants' new home city (Appadurai, 1996).

A second modality of transnational politics involves states seeking to enlist their diasporic communities into the process of national, regional and even local economic development. Such 'diaspora strategies' (Larner, 2007: 332) have been employed by states as diverse as India, Honduras, Mexico, Ireland and New Zealand with the intention of mobilising a nation's (and in some cases a region, or city, or village's) non-resident citizens (and also often overseas born co-ethnics) in aid of national economic development. These strategies reconfigure the spatial reach of the state in all sorts of interesting ways. They also create sets of institutional interdependencies that weave the spatiality of a nation's transnational diaspora into its existing territorial structure. Which brings us to the third modality through which transnational political action is reshaping cities. This is the way in which the circulation of people, ideas, information, money and goods that accompanies the emergence of transnational social morphologies alters the horizons of possibility in the places that it encompasses. The vision of moving somewhere else, or back to some place, or an experience of a certain kind of life, or even the idea of a certain life, in many instances profoundly alters the shape and dynamics of a place's political culture. Thus, to take one small example, in the Pacific Kingdom of Tonga overseas educated Tongans have taken the lead in calling for greater democracy in the kingdom. But in November 2006 when pro-democracy activists took to the streets of the capital Nuku'alofa their protests rapidly descended into a chaos of arson and looting as gangs of youths, many of whom had grown up in Auckland, Sydney, or Salt Lake City and been sent back to Tonga by their families to protect them from lives of delinquency, took things into their own hands. It is just such strange and sometimes disturbing phenomenon that transnational urbanism seeks to make sense of.

51

## KEY POINTS

- Cities are defined in all sorts of ways by their connections with distant places.
- Globalisation is not just the product of large, powerful and highly visible actors, disadvantaged and seemingly powerless individuals and groups also produce it from below.
- Transnationalism is a concept that seeks to make globalisation conceptually manageable. It focuses on the concrete practices that allow

social, economic and political relationships to be stretched across national borders.

- Trans-migrants of all social backgrounds construct intricate social, economic and political networks that knit together places over often very substantial distances.
- These transnational social morphologies profoundly shape the dynamics of the cities that they encompass.

## FURTHER READING

Michael Peter Smith's (2001) *Transnational Urbanism: Locating Globalization* provides a comprehensive overview of the key literatures on globalisation and cities, and provides a compelling argument for the need to study the transnational networks created by migrants. Peter Jackson, Phil Crang and Claire Dwyer's (eds) (2004) *Transnational Spaces* is an excellent and wide ranging collection of essays that shows how geographers have used the concept of transnationalism. In 'Transnational urbanism: everyday practices and mobilities', a special edition of *Journal of Ethnic and Migration Studies,* David Conradson and Alan Latham (eds) (2005a) present a series of articles addressing the everyday practices through which transnational migrants organise their mobility. It includes a short introductory essay by Michael Peter Smith.

*AL*

# 2 Constructions

# 2.1 NATURE

At first glance the concept of nature (and of the natural) seems to suggest something opposed diametrically to the urban: if the former conjures up a vision of untrammelled, primal wildness untouched by human agency, the latter connotes all that is modern, artificial, and socio-technically constructed. As such, the urban would seem to be defined in terms of what the natural is *not*. But such strict opposition is far too simplistic, not least because it is often difficult to delimit the imaginative and material boundaries between what is understood as natural and what is understood as urban.

As the cultural theorist Raymond Williams (1976; see also 1973) observed, nature is one of the most complicated, contested and dynamic of concepts. Williams' discussion provides a useful framework within which to begin thinking about the relation between nature and urban geography. To begin, the nature of the city can be taken to refer to the 'essential quality and character *of*' the urban (ibid. p. 219). Much of the history of urban geography can be interpreted in this light: as an attempt to define the essential processes and patterns of cities. Then, and second, nature can be understood in rather more metaphysical terms 'as the inherent force which directs either the world or human beings or both' (ibid. p. 219). Even if few contemporary scholars of the urban would hold such a view, the idea that cities and their inhabitants are shaped and organised by forces operating behind or beneath the surface of everyday life remains an influential one – underpinning the idea that urban society is the spatial expression of the logic of capital accumulation, for instance. Third, and finally, nature can be employed to refer to 'the material world itself, taken as including or not including human beings' (ibid. p. 219). The degree to which humans are understood as part of this natural world is important. If humans are understood to be separate from the natural world, then cities will tend to be viewed (often critically) as symptomatic of how far modern societies have become alienated from nature. But if considered part of nature, then contemporary urban life cannot be ontologically distinct from the less urbanised environment with which it is surrounded. Rather, cities are manifestations of different kinds of nature – or natures – than rural or non-urban spaces.

# Imagining nature in the city

One way to begin working through the complex relation between ideas about nature and theories of the city is to consider how such ideas have influenced a specific set of theories about what cities are and how they work. An instructive point of departure in this regard is the Chicago School of urban sociology, the chief proponents of which included Robert Park, Ernest W. Burgess, Homer Hoyt, Chauncy Harris and Edward Ullman. One of the most distinctive elements of the work of these figures was the degree to which it was informed directly by ideas from biology, ecology and from studies of evolution. The assumption underlying this work is that even if they are social and cultural, cities are shaped by processes analogous to those observable in nature. This way of understanding cities is most evident in the kinds of analytical metaphors employed to describe fundamental units and processes of urban space. So, for instance, the city could be understood as a kind of 'social organism', composed of 'natural areas' or 'ecological units', each of which was characterised by a distinctive and relatively stable mix of inhabitants. And, in turn, transformations in these areas could also be understood via concepts such as 'invasion' and 'succession'.

The writing of Ernest Burgess is particularly illustrative in this regard. For Burgess, the internal growth and transformation of individual cities could be understood as the consequence of processes of 'organisation and disorganisation analogous to anabolic and katabolic processes of metabolism in the body'. This kind of metaphorical description raised a set of important questions about what exactly the 'normal' or 'natural' rate of city growth was and about how immigrants were incorporated into and became 'organic' elements of the life of the city. It also facilitated the measurement of 'disturbances of metabolism caused by excessive increase' (quoted in Miles et al., 2000: 25). Yet while this conceptual vocabulary can appear to lend scientific credibility to theories about how cities work, it is important to remember how it is also constructive of a particular imaginative geography of urban space. Thus, the same vocabulary also allows Burgess to speak of 'racial temperament' and to give a normative and moralising slant to the movement and mobility of people insofar as they disrupt the 'organic' relations within cities. As this suggests defining cities in terms of 'natural processes' may be generative of a conceptual discourse with problematic political resonances.

55

Ideas about the nature of cities not only inform academic theories. They have also influenced attempts to plan, design and engineer urban space. The dramatic acceleration in the rate of industrialisation and urbanisation during the nineteenth century precipitated pressures for the reform and redesign of urban space. These pressures emerged in part through a concern with the deleterious physical effects of urban spaces increasingly devoid of any natural elements: but they were also borne of a concern with the moral degradation and psychological impoverishment associated with urban environments. Municipal parks are the most visible expression of how ideas about the physical, social and moral benefits of contact with nature were translated into the design and transformation of the city. One of the more famous expressions of the attempt to plan and design urban nature can be found in Central Park in New York, construction of which began in 1858 under the supervision of the architect Frederick Law Olmstead. By combining elements of the pastoral landscape (sweeping meadows in the southern part of the park) and the picturesque (rocky, irregular topography in the north) Olmstead's aim was to provide the inhabitants of New York with access to nature without having to leave Manhattan. This process of landscape design did not come without its costs, necessitating, for instance, the clearing of slum areas occupying the site of the planned park (See Katz and Kirby, 1991).

The question of how to design cities in order to best incorporate the elemental benefits of nature continued to be central to urban planning and design in the twentieth century. While very different, the respective utopian visions of architects like Le Corbusier and Frank Lloyd Wright were both concerned with engineering urban space in ways that offered new possibilities for integrating the natural and the artificial. Le Corbusier's 1922 *Ville Contemporaine* proposed a vision of a modernist machine for living that incorporated well-ordered and regimented green spaces. In contrast, Frank Lloyd Wright's *Broadacre City* – was a vision of the low-density dispersal of the city into the country, one that anticipated the layout of the North American suburban landscape. In the UK, the ideas of Ebenezer Howard influenced the emergence of the idea of the Garden City Movement, the legacy of which is visible today in Letchworth and Welwyn Garden City. Howard's writing reiterates the renewed emphasis on the role that nature should play in sustaining the life of modern cities. For Howard, writing in 1898, 'neither the town magnet nor the country magnet represents the full plan and purpose of nature. Human society and the beauty of nature are meant to be enjoyed

**Figure 2.1.1** 'Madinat Jumeirah in Dubai: the design of an idealised urban nature is a key part of contemporary urban megaprojects

together. The two magnets must be made one. As man and woman ... supplement each other, so should town and country' (Howard, in Kaika, 2004: 18). The key point here is that planning and designing cities is a process involving the production and reproduction of particular kinds of imaginative geographies about the relation between nature and city.

## Material urban natures

Urban natures are imaginative: but they are also profoundly material involving the ordering, circulation and manipulation of things like food, building materials and electricity. When considering the **materiality** of urban natures, two traditions of thinking are particularly important. The first is a Marxian-influenced critique of political economy in which cities are understood as sites for the conversion of what Neil Smith (1984) calls first nature (nature untrammelled by human activity) into second nature (nature transformed, worked upon, manipulated, com-modified) in a process that generates radically uneven and contested geographies (see also Davis, 1990). Drawing loosely upon this tradition, environmental historian William Cronon has illustrated how the city of Chicago was drawn into a complex and dynamic web of spatio-temporal connections with its hinterland such that 'first and second nature mingled to form a single world' (1991: 93). Transport technologies, most notably the railway, were central to this process. As Cronon puts it: in the second half of the nineteenth century, city and country, linked by the '"wild scream of the locomotive"' would together work profound transformations on the western landscape' (1991: 93).

A second and related set of intellectual resources for thinking about the materiality of urban natures is concerned with amplifying its hybrid quality: that is, with the way in which the nature of urban space is a kind of co-fabrication between human and non-human agencies. Here, geographers and others have drawn upon the work of thinkers such as including Donna Haraway and Bruno Latour. In Donna Haraway's (1991) terms, cities can be understood as cyborg environments – complex assemblages of the natural, the technological and the social (see also Luke, 1997). Describing the nature of the city in these terms conjures up a dystopic vision of the urban: the kind of vision realised in Fritz Lang's *Metropolis* (1929) or Ridley Scott's *Bladerunner* (1980). But the hybrid or cyborg natures of cities are rather more mundane than these cinematic renderings suggest. They can also be revealed through the kinds of technical **infrastructures** required to move various 'natural' elements around cities. Water has received particular attention in this respect, perhaps because its fluidity disturbs conventional assumptions about the objectness of urban **materiality**. As a number of geographers have demonstrated, tracing the networks and practices implicated in the treatment, transport and consumption of water reveals the city as a hybrid space, in which purity of water is a socio-technical product (see Gandy, 1997, 1999; Swyngedouw, 1997). This work attempts to hold onto the fundamental insights of political economy while also rendering it more attentive to a range of hybrid agencies and technologies. In doing so, they return to the kind of metaphors employed by Burgess in the 1920s, although to very different ends, and with rather different socio-political leanings. As Maria Kaika argues, 'the myriad of transformations and metabolisms that support and maintain urban life, such as water, food, computers, or movies always combine environmental and social processes as infinitely interconnected' (2004: 22).

## 'More-than-human' urban geographies

Whether or not we choose to return to organic metaphors of metabolism, the work outlined above reveals how the mobilisation and movement of objects, fluids and things ties the urban and the natural into dense networks of material connectivity. At the same time it also draws our attention to how other forms of life participate in the production, transformation and consumption of urban natures. Animals are particularly important here insofar as they complicate the conceptual and material distinction between the natural and the urban. Cities are sites in which the bodies of animals become part of the metabolic processes of which urban natures consist. Set in Chicago, Upton Sinclair's (1905) novel *The Jungle* contains the most vivid depiction of how the modern city functions in this regard, illustrating the role animal bodies play in the broader geographical processes described above by William Cronon: **59**

> There was a long line of hogs, with squeals and life-blood ebbing away together, until at last each started again, and vanished with a splash into a huge vat of boiling water. It was all so very businesslike that one watched it fascinated. It was pork-making by machinery, pork-making by applied mathematics. [...]. The carcass hog was scooped out of the vat by machinery, and then it fell to the second floor, passing on the way through a wonderful machine with numerous scrapers, which adjusted themselves to the size and shape of the animal and sent it out at the other end with nearly all of its bristles removed. (Sinclair, 1905: 40)

The changing industrial and economic geography of western cities means it is unlikely that such slaughterhouses would now be located within the centre of a city like Chicago. But animals remain part of the diverse natures of which cities are composed (Wolch et al., 1995; Philo, 1995). Through domestication they have become an integral part of the social, emotional and economic life (Anderson, 1995): at the same time, a range of semi-wild or feral species have begun to become more visible in cities generating new kinds of urban ecological niches (Hinchliffe, 1999). Additionally, the presence of zoos illustrates how the natures on display in the city are often the product of geographical and historically situated technologies of classification, capture and containment, through which the exotic animal other is spectacularised for the pleasure of an urban audience: an 'illusion of Nature [...] created from scratch and re-presented back to human audiences in a cultural performance' (Anderson, 1995: 275).

## Mutliplying urban ecologies

While the conceptual vocabulary employed by geographers to comment critically upon nature has become increasingly sophisticated, it would be a mistake to reduce this commentary to a narrative of critical disenchantment. It remains possible to affirm that which we call nature as a set of hybrid processes through which to cultivate benign forms of attachment and involvement in the city. This does not automatically mean conjuring up strange new worlds. It can also mean, for instance, generating affective investment at and through particular sites of practical and embodied activity. Indeed it is through the diverse energies and activities of the **body** that the processual dimension of urban natures can be most easily apprehended. For instance, a whole range of mundane and microscale opportunities exists for modes of corporeal engagement and experiment with distinctive, and more-than-human ecologies in the city (Hinchliffe et al., 2005). While such opportunities are afforded by large-scale and relatively intensively managed spaces such as Central Park, they can also be apprehended in interstitial zones – sites of ruin, dereliction, or decay – whose neglect is potentially generative of benign contact zones in the city. The important point is to find opportunities for the cultivation of an ethics and politics of cautious enchantment through the proliferation of urban natures (Bennett, 2001).

Have we therefore arrived at the end of a conception of nature as a space defined as distinct from the urban? If we cleave to a single understanding of the concept of nature, one defined in terms of essential ontological purity, then the answer to this question is yes. Yet rather than an occasion for nostalgia, this might better be understood as an opportunity. As Neil Smith argued more than 20 years ago, to oppose nature to society or the urban no longer makes any sense. And if, like Raymond Williams, we understand nature to be a dynamic concept whose meaning has varied historically, geographically, and socio-culturally, then it remains important to think about the many kinds of nature(s) that are entangled and manifest in urban life as part of a wider project attentive to the spatial politics of hybrid geographies (Whatmore, 2002; Hinchcliffe et al., 2005).

## KEY POINTS

- The concept of nature has multiple meanings.
- Different ideas about nature have shaped *both* ideas and theories about cities *and* attempts to redesign the physical spaces of urban life.
- Nature is no longer understood by geographers as ontologically pure: rather, it tends to be conceived in terms of a set of hybrid relations between physical and imaginative, human and non-human processes.
- Cities are therefore not defined against nature, but are locations where hybrid natures are produced and transformed with particular intensity and complexity.

## FURTHER READING    61

Raymond Williams' (1973) *The Country and the City* provides a useful point of departure for thinking about the imaginative relation between the urban and the rural. *William Cronon's (1991) Nature's metropolis: Chicago and the Great West* is a highly readable narrative about the complex relations between the development of Chicago and the transformation of its hinterland. Matthew Gandy's (2002) *Concrete and Clay: Reworking Nature in New York City* offers an empirically rich account of the imaginative and material reworking of nature in a global city. In a similar vein, Maria Kaika's (2005) *City of Flows* considers the complex participation of water in urban planning and visions of the modern city. For an overview of the place of animals in urban life see Jennifer Wolch's (2002) discussion of: *'Anima Urbis'*. For a provocative vision of how we might reimagine the conceptual and political terms through which we Think about urban natures, see the discussion of *'Urban Wild Things'* by Steve Hinchliffe, Matthew Kearns, Monica Degen and Sarah Whatmore (2005).

*DMcC*

# 2.2 MATERIALITY

In its most crude sense the materiality of the city is the stuff of which the urban environment consists. Taken literally, the materiality of the city refers to the physical things, objects and structures that give urban space its shape and substance: those things encountered and used in myriad ways as part of the everyday lives of urban dwellers. There is something reassuringly concrete about this definition of materiality: it seems to refer directly to that which is obviously visible, undeniably tangible and demonstrably durable about cities.

Yet materiality does not just designate brute matter. For one thing, this is because the things we encounter and use have a certain symbolic value – think for instance of the many commodities used in urban life. In many cases their value has as much to do with what they symbolise as it does with their physical quality: clothing and footwear are obvious cases in point. Furthermore, there are many phenomena that seem to shape our experience of urban life without necessarily being physically tangible – dreams, imagination, desire, emotions. What kinds of materiality might characterise such phenomena? Finally, the idea that materiality refers to physical objects is complicated by the fact that much of what allows cities to work does not take this form – water, electricity, light and heat have a material quality that cannot be reduced to the shape of a thing. So any account of the materiality of urban geography must involve more than an inventory of its obvious physicality materiality needs to be understood also as a spatio-temporal *process* in which the more tangible, physical stuff of the city is a lively participant.

## Questioning materiality

It is fair to say that materiality has been relatively under-conceptualised within geographical approaches to urban space, and indeed within the discipline more widely. On one level this claim might appear surprising: after all, geography is a discipline that places particular emphasis on the study of the relation between human activity and the physical environment. Furthermore, a central thread in the emergence and consolidation of geography as a discipline worthy of the name has been a concern with the human shaping of the material landscape. Such work is exemplified most famously in the work of Carl Sauer (1925), for whom it was possible to trace the diffusion of cultural activity through the distribution of certain kinds of material things including buildings, tools and agricultural forms. While Sauer himself was not an urban geographer – indeed he was distinctively anti-urban in his thinking – his understanding of landscape as a material form whose morphology is transformed over time through the agency of human activity can easily be applied to urban landscapes. The material form of cities can be read and interpreted as evidence of the kind of cultures through which they have been shaped.

Clearly there is something to this. The material form of architectural styles and patterns of building construction can be interpreted as evidence for the presence and diffusion of various cultural practices. Why then question what seems like a common-sense conception of materiality? There are at least three sets of reasons why we might do this. The first revolves around questions of ontology: that is, around questions about the fundamental nature of materiality, about its essential qualities. While admittedly concerned with change over time, in the work of Carl Sauer materiality tends to be understood as a kind of substance whose nature does not change even if its form or shape does. This then is a rather static conception of materiality, in which transformation is really only a change in the superficial organisation, structure and appearance of something more essential. Put another way, materiality is rendered inert, reduced to brute matter. Yet, as geographers and others have argued, the stuff of materiality is dynamic, as are theories about what constitutes matter (Kearns, 2003).

A second set of reasons for questioning revolves around the relation between materiality and those phenomena and processes sometimes called 'immaterial'. This category includes various kinds of representation including images, sounds and signs; various kinds of experience such as dreams, desires and memories; and a range of qualities of urban experience that seem difficult to define while also registering something that makes a degree of intuitive sense – this category includes mood, atmosphere, spirit. It is not unfair to say that for much of its history, the discipline of geography has struggled to come to terms with these phenomena and their relation to material landscapes. Consequently, much of geographical knowledge has come to rest upon an implicit and under-theorised divide between the material and the immaterial, with

63

the former understood as something that can only be mapped fully by the knowing, rational and effectively dematerialised (because disembodied) subject. But this division is increasingly difficult to sustain. As a result, urban geographers are rethinking the relation between the materiality of urban spaces and a range of apparently 'immaterial processes' (Latham and McCormack, 2004).

A third and final set of reasons for questioning materiality revolves around issues of agency (who, or what can act) and power (the capacity to affect the actions of other agencies and, to a certain extent, to be affected by the activity of those agencies). As geographers and others have argued, access to and control over material resources is by no means equally distributed, and is also a crucial factor influencing the physical form and structure of urban space. So any conception of materiality needs to have a sense of how the physical capacity to make and unmake things (see Graham, 2004b) and, perhaps equally importantly, to make things matter, are crucial issues. But this claim again raises all sorts of questions. Does power reside in the capacity of certain individuals or agencies to exercise control over material resources? Alternately, is materiality the resultant *effect* of the capacity of various agencies? Perhaps even more provocatively, does the materiality of non-human forces and agencies have a power in itself, one not dependent upon its mobilisation by human activity?

# Re-materialising urban geography?

Each of these areas of questioning might appear rather abstract insofar as they move away from what seems most taken-for-granted about the materiality of urban space. But they form the implicit background for the efforts of urban geographers and others to engage with and think through the materiality of cities over the last few decades. These efforts can be differentiated along the lines of a number of ongoing research trajectories.

## Materialising capital

The first trajectory revolves around a concern with the dynamic processes through which the built environment is materially produced and, equally importantly, the contested and politicised nature of this process. This line of research emerges with greatest force from geographers' reading of the historical materialism of Karl Marx and its subsequent reworking by various Marxist thinkers. The legacy of Marx's thought is obviously complex, but for the purpose of the present discussion its significance rests upon the claim that the analysis of modes of social and cultural production cannot be divorced from an understanding of material processes, the most obvious of which is the necessity for humans to transform their environment in order to survive. The relations of which social and economic life consist are, for Marx, always embedded in the process of this material transformation. Crucially, as David Harvey demonstrated in a series of key works from the early 1970s into the 1980s (1973, 1982, 1985), these relations have a spatial dimension in the sense that they are productive of actual landscapes. Thus, under capitalism, the urban built environment becomes a crucial part of the process of generating profit (or loss) through circuits of capital accumulation. As Harvey put it in *Social Justice and the City*:

> Under capitalism there is a perpetual struggle in which capital builds a physical landscape appropriate to its own condition at a particular moment in time, only to destroy it, usually in the course of a crisis, at a subsequent point in time. The temporal and geographical ebb and flow of investment in the built environment can be understood only in the terms of such a process. The effects of the internal contradictions of capitalism, when projected into the specific context of fixed and immobile investment in the built environment, are thus writ large in the historical geography of the landscape that results. (1973: 124)

For Harvey then, as for other geographers including Neil Smith (1979) and Allen Scott (1980), the physical form of urban – and sub-urban space – becomes a spatial materialisation of historically and geographically specific circuits of capital. In turn, this emphasis on how capital materially shapes the built form of the urban landscape has been modified by efforts to foreground urban space as a site of material and symbolic consumption. These modifications reflected the influence of cultural theorists (such as Raymond Williams) and of the 'new cultural geography' (exemplified in the work of Stephen Daniels, Denis Cosgrove and Peter Jackson) that emerged during the 1980s and 1990s, and which emphasised the complex relation between material landscapes and representational processes. The influence of this greater attention to representational meaning can be discerned in arguments about the emergence and 'restlessness' of postmodern urban landscapes (see Knox, 1991; see also Gottdiener, 1994).

Yet for many geographers interested in the analysis of urban environments within the critical tradition influenced by Marx's historical materialism, too much emphasis on representational processes could produce a rather impoverished or 'ungrounded' sense of materiality. Fears about the critical and political costs of an increased focus on the representational have recently prompted calls for a cautious 're-materialisation' of urban geography (Lees, 2002) – calls that echo moves in the wider discipline (see Jackson, 2000; Matless, 2000). Yet, lest we think that such calls offer a return to a simpler, more concrete version of materiality, it needs to be remembered that at the heart of urban political economy are things whose nature is far from concrete – commodities. Indeed, as Marx was all too aware, explanations of the nature of the commodity often requires recourse to immaterial and abstract forces.

## Material urban cultures and practices

A second set of ways in which geographers are addressing some of the questions outlined in the previous section revolves around investigating how urban cultures are produced and consumed through a range of material practices involving things and objects. This is also perhaps where the concerns of geographers overlap most directly with a series of other disciplines, notably cultural studies, archaeology and anthropology (see, for example, Appadurai, 1986; Attfield, 2000; Miller, 2005). Geographical studies have revealed a variety of ways in which material cultures are produced and reproduced through the design, use and spatio-temporal biographies of things. At the heart of such studies is the argument that materiality is an active participant in how the experience of urban space is made meaningful. This meaning is lived and negotiated through material practices; including eating (Cook and Harrison, 2003), urban regeneration (Atkinson, 2007) and tourism (see Haldrup and Larson, 2006). For instance, we can think of the importance of urban gardens to the performative maintenance of a sense of self, identity and well-being. These green spaces are cultivated and managed through a range of tactile, physical practices (digging, weeding, trimming, mowing, watering) and with the aid of various materials (tools, seeds, water, fertiliser, weed killer, garden furniture). Yet, as Hitchings (2007) has argued, such materials are never simply 'under control': rather, they have an unpredictable – and only partially manageable – nature, excessive of the intentional action of human agency. Put another way, even where it is intimately intertwined with a human sense of place, the materiality of material culture is badly behaved.

66

The excessive character of material culture is important in another obvious respect. This concerns the sheer volume of 'stuff' of which urban environments are generative. In turn, this points to a further inflection of the term materiality – its relation with the claim that contemporary industrial societies have become increasingly 'materialistic', to the extent that they have a 'throwaway' attitude to things and objects. As Gregson, Metcalfe and Crewe (2007: 698) have argued, however widely held, such claims do not necessarily bear much scrutiny when we examine how people deal with the excess of material urban culture (see also Hetherington, 2004). Thus, the disposal of 'consumer goods is a long way from a matter of automatic waste generation; [...] such acts frequently involve saving as well as wasting things; and, critically, [...] the process of discarding is one of the key ways in which we make present and materialise some of our primary social identities and the love relations that sustain them' (Gregson et al., 2007).

## Relational urban materialities

The practices and objects outlined above obviously never relational in isolation: instead, they are better conceived as part of complex ecologies. Geographers have addressed the relational dimension of urban materiality in part by drawing upon elements of Marxian thought outlined above. At the same time these ideas have been modified by a concerted attempt to conceptualise materiality as a co-production between a range of human and non-human agencies. Important influences on this modification are the set of ideas known as actor-network theory, often associated with thinkers including Michel Callon, Bruno Latour and John Law; and the writing of the feminist scholar of technoscience Donna Haraway, who uses the figure of the cyborg as a way to conceptualise how science and technologies produce realities that are at once material and semiotic. Materiality in these terms is less a fixed entity than a set of hybrid and relational processes that do not obey the neat ontological divide between subject and object, nature and culture. The emphasis here is on tracing how specific urban geographies of materiality emerge through complex networks of circulation, distribution and connection: water and electricity are two of the most obvious examples (see the entry on **nature** and **infrastructure**; see Swyngedouw, 1997; Kaika, 2004).

A further important element of this emphasis on relational materiality concerns the question of technology and the degree to which it is implicated in the transformation of materiality. Rather than something which acts upon an inert material world, technology always operates within a world in which the '"material" and the "social" are always already bound together, always already binding together' (Bingham, 1996: 635). Think, for instance, of the construction of familiar things including as roads, buildings, or various infrastructural networks. If we think of these constructions as evidence of how the social transforms the material then we surely miss something of the dynamic processuality through which materiality comes to matter (Latham and McCormack, 2004). Such identifiable and familiar urban forms as roads are transformative co-constructions drawing together diverse sets of agencies.

Furthermore, the design of certain technologies reveals that materiality is neither singular nor homogeneous: it has multiple registers, each of which is transforming. Indeed, according to some geographers it is possible to discern the emergence of distinctive and novel registers of materiality in urban life. Nigel Thrift (2005a), for instance, argues that such emergence is taking shape around the inventive, infrasensible materialities of screen, code and body. Of these, code is of particular interest to urban geographers because it is thoroughly embedded in transformations in spatial relations within and between cities, at the same time as it facilitates movement between different material registers (for instance between the body and the screen) (see also Dodge and Kitchen, 2005). Rather than trying to work out exactly what kind of object code actually is, it might therefore be more profitable to try to understand how it facilitates the recombinant emergence of what Adrian Mackenzie calls a 'transductive field' (2002), from which various materialities precipitate.

## Materiality in process

Any use of materiality as a concept within studies of urban space needs to be accompanied by an acknowledgement that, materiality is not reducible to or coterminous with the object – otherwise it would be difficult to account for the real, tangible force of events and experiences that lack object-like status. In contrast, materiality is best thought of as a process from which identifiably discrete objects may emerge to take on a certain degree of consistency. This emergence is often shaped by certain kinds of constraints, deliberate interventions and recalcitrant agencies.

To make these qualifications does not diminish the sense in which materiality is entangled in questions of power. Indeed, it arguably allows urban geographers to work towards a more nuanced understanding of the relations between cities, power and materiality. As such, any attempt to 'rematerialise' urban geography cannot simply be a return to some already existing, solid, concrete sense of the material. Materiality is always in process, even if this process unfolds at different speeds.

## KEY POINTS

- Materiality has often been under-conceptualised within urban geography.
- Any definition of materiality must not reduce it to brute matter, but recognise that there are many ways in which it can be theorised.
- Urban geographers and others are beginning to develop an understanding of materiality as a dynamic process rather than a static thing.

## FURTHER READING
69

Loretta Lees (2002) paper, *Rematerializing Geography: The 'new' urban geography* provides an example of recent efforts by urban geographers to reclaim the material in the wake of what are sometimes perceived as an excessive concern with an 'immaterial' realms of images, texts and representations. Alan Latham and Derek McCormack's (2004) paper *Moving cities: Rethinking the materialities of urban geographies* builds upon and responds to Lees' argument by outlining a number of empirical and conceptual pathways into how geographers might rethink the materiality of urban geography. In doing so they complicate any attempt to invoke the material in ways that juxtapose it to the immaterial. A good overview of these debates and their wider context is contained in Phil Hubbard's book (2006) *City*. Other geographers have focused on the materials cultures of cities. The work of Russell Hitchings (2007) provides an interesting example of how geographers are rethinking the nature of urban material cultures: in this case through research into urban gardening.

*DMcC*

# 2.3 INFRASTRUCTURE

The ability to construct and provide infrastructure – from motorways to IT cables to electricity and gas pipes to water and sewerage systems – bears a close relation to the economic prosperity of a particular place. It requires a huge amount of investment, given the difficulty of land purchase, excavation, materials, construction and maintenance, costs that few private firms are willing to undertake. Such 'megaprojects' as the installation of a major sewer system, motorway, or power station thus tend to be provided by the state, through public funds. The story of infrastructure development is one of great significance in the development of the contemporary city (Graham and Marvin, 2001), but also in histories of the nation-state and international systems of trade and interdependency. For geographers, one of the most interesting aspects of infrastructure is the distanciation it affords. The relationships often associated with technological change, from the telephone (an early facilitator of long distance friendship), to the internet (now heralded as completely deconstructing established ideas of public and private, home and the nature of community itself) are seen as drivers of socio-spatial change.

70

## Cities, infrastructures and the nation-state

Often conceived as an object of great pride by prime ministers, military leaders, or boosterist regional politicians, and underpinned by a popular culture of travel and mobility (Urry, 2000; Cresswell, 2006), motorways, railways, airports and bridges have long had a powerful hold on urbanised national territorial imaginings. In addition to these innovations, which allow for the faster movement of people, there is also the case of infrastructures which move things, such as water and oil pipelines, dams, telephone and electricity networks and – most recently – internet cables. Long associated with discourses of modernity and modernisation, these technological developments have replaced a traditional valorisation of nature as the source of awe and wonder (Nye, 1996). Huge megastructures such as the Hoover Dam in the US or the Three Gorges Dam in China have celebrated the human domination of nature, a central feature of the politics of modern states.

**Figure 2.3.1** 'Golden Gate Bridge, San Francisco. The construction of such "technological wonders" (Nye, 1996) has been a major part of the urban experience'

These are of major concern to geographers, as infrastructure effectively shrinks distance, compressing time and space (Harvey, 1989). Given recent concerns about the arrival of a new phase of globalisation, it is also important to consider how by shrinking distance, nation-states can be seen as territorially more cohesive. Airports are a case in point. While often associated as a facilitator of globalisation, airports are often seen in geopolitical terms as much as a means of uniting large dispersed national territories, as of engaging with extraterritorial actors. In Malaysia, for example, the new Kuala Lumpur International Airport was sited in Sepang, at the centre of an emerging high technology corridor that sought – along with a repositioning in the regionalised global economy – to form a '"backbone" controlling the national geo-body' (Bunnell, 2004a: 109). In the United States, the construction of a federal system of airports was a fundamental part of the country's post-war development and growth. Motorway systems 'link spaces together via car driving … [which includes] stitching the nation together. The "democracy" of car travel enables valorised national scenes and sites to be visited, opens up the possibilities for "knowing" the nation … Previously remote places can be reclaimed within a national geography' (Edensor, 2002: 127).

However, infrastructure development can also be divisive. The provision and distribution of water has always been a key aspect in the development of urbanised nation-states. A good example is provided by Swyngedouw (2007), who discusses the role of water infrastructure in the context of Francoist Spain. After the Civil War of the 1930s, the construction of national hydraulic plans was a keystone of the Francoist dictatorship's nation-building project, with Spain having the highest number of dams per capita in the world: 'Water issues were constructed as the main collective challenge facing Spain, thereby deflecting attention from issues like social justice or land distribution' (p. 12). Furthermore, the construction of this infrastructure was made possible through 'networks of landowners, large industrialists, engineers and media', which 'produced a unitary national territorial complex, and eliminated dissenting political voices, regionalist impulses, and alternative configurations' (p. 24). The interesting point that emerges from this discussion is its 'technonatural' dimension, in that an urbanised nation-state rests upon a fundamentally 'natural' resource whose infrastructures may well be invisible to urban dwellers.

The construction of rail networks is another fascinating area of debate. For example, the opening of the Channel rail tunnel between Britain and the European mainland served to destabilise long-established discourses of national identity. As Eve Darian-Smith (1999: 81–2) has argued in a notable legal anthropology: 'The engineering and technological debates surrounding the viability of a cross-Channel tunnel or bridge were of considerably less importance than the mental anxieties and competing moral dilemmas that the Tunnel has over time symbolised'. (Darian-Smith, 1999: 81–2). Here, the deeply held idea of London at the apex of a national hierarchy with Kent as an idealised countryside (the 'garden of England') suggests that the Tunnel represents a respatialisation of London and Britain away from a clear territorial commonsense, tying London into a Brussels–Paris nexus and opening up new regional spaces. Thus, the Tunnel represents a number of cultural and psychological transformations which deconstruct the fixed essences of region, city and nation (Sparke, 2000).

However, it is true that some places – often city-states such as Singapore, Hong Kong and Dubai – position themselves and their nation-building strategies in the context of global flows of passengers and goods through open skies policies, or as 'free trade' zones in the midst of major regional markets such as Southeast Asia or the Middle East. Airports, high-speed rail links, dams, tunnels, motorways and large mixed use developments are defining features of a world increasingly dependent on rapid transport and infrastructural improvement. In every major regional economy, whether in the rapid modernisation of infrastructure seen in Pacific Rim economies (Olds, 2001) or in Latin America, to the construction of trans-European networks as a central plank in European Union integration, to the continuing attempts to modernise the old industrial cities of the US and Canada, arguments for spending on major projects are widespread.

73

# Splintering urbanism

Infrastructure thus plays a key role in the development of a relational human geography, one that emphasises a metaphysics of fixity and flow. It feeds into the debate on globalisation: 'Relations between people, firms, institutions, communities and buildings on the global scale ... may in many cases be more significant than their relations with urban activities or spaces that are physically adjacent' (Graham and Marvin, 2001: 206). As a result of this, 'Places are not contiguous zones on two-dimensional maps' (p. 203). This means that rather than understanding the composition of different places through solely look-ing at a map, they need to be seen as having a spatio-temporal dimension. This fits in with ideas of relational geographers, who understand cities by the intensity of their interconnectivity.

Graham and Marvin's *Splintering Urbanism* (2001) synthesises many of these trends. This rather inviting metaphorical take on the contemporary city is based on four pillars (2001: 10–12). First, they see cities as 'sociotechnical processes', where everyday practices such as washing, cooking, going to the toilet, putting out rubbish, etc., are made possible by a range of technical practices so mundane as to be taken for granted in the west. Bodies and machines (however crude) interact in a range of time-saving and comfort-giving interplays. Second, these are based upon a set of material infrastructures – sewers, pipes, cables, aerials, roads – which link people and things together. Such networks 'unevenly bind spaces together across cities, regions, nations and international boundaries while helping also to define the material and social dynamics, and divisions, within and between urban spaces' (p. 11). At its extreme, this helps to explain processes of globalisation, where distance – especially for communicative practices – is diminished markedly. Third, such networks are costly to build and maintain, involve large capital investments, and 'represent long-term accumulations of finance, technology, know-how, and organisational and geopolitical power' (p. 12). Fourth, networked technologies such as 'heat, power, water, light, speed and communication' (p. 12) are fundamental to understanding the contemporary urban experience, its sensory conditions, its visual conditions and sense of spectacle; in short, they contribute fundamentally to urban culture.

The intensification of global business travel has facilitated the development of an international class of technical experts such as civil engineers, architects, project managers and cost consultants who are involved in the construction of megaproject developments (Olds, 2001). Yet the costs of these often ambitious projects can be unpalatable for politicians as they seek to control public spending. In 1980, Peter Hall's bluntly – and memorably – entitled *Great Planning Disasters* described a range of major public sector calamities: the Anglo-French Concorde, Sydney's Opera House, San Francisco's BART system, London's third airport and motorways. In each of these case studies, Hall pointed the finger at a range of private sector interest groups, politicians and bureaucrats, who for a diverse and usually rather Machiavellian range of reasons chose to underestimate or ignore costs, or provide a politicised method of demand forecasting. As Flyvbjerg, Bruzelius and Rothengatter argue in their book *Megaprojects and Risk* (2003), 'at the same time as many more and much larger infrastructure projects are being proposed and built around the world, it is becoming clear that many such projects have strikingly poor performance records in terms of economy, environment and public support' (p. 3). The costs of major infrastructure projects are thus likely to be contested and governments are seeking new ways to finance large public projects.

It can be argued, therefore, that the growing sense of stress placed upon urban infrastructure has coincided with a crisis in public will to pay for them. A widespread shift away from public collective provision has meant that only privatised solutions are entertained, which in turn has involved a retreat from the modernist ideal of universal access to publicly distributed services. In the last half of the nineteenth century and early decades of the twentieth century, a 'growing corps of urban engineers sought to understand the growing industrial city as a systemic "machine" that needed to be rationally organised as a unitary "thing", using the latest scientific and technological practices available' (p. 44). This 'unitary city ideal' was premised upon a model of balanced, evenly spread service provision with universal access notion whereby 'comprehensive, integrated networks of streets could be laid across whole urban areas in a technocratic way, to bind the metropolis into a functioning "machine" or "organism"' (Graham and Marvin, 2001: 53). Underpinned by electrification, and with the laying out of regularised streets on grid systems, infrastructure provision was fully integrated into the urban planning process. By the 1960s, however, this 'integrated ideal' was being challenged, public infrastructure entering into crisis. There is now a new orthodoxy in many branches of urban planning: 'The logic is now for planners to fight for the best possible networked infrastructures for their specialised district, in partnership with (often privatised and internationalised network) operators, rather than seeking to orchestrate how networks roll out through the city as a whole' (Graham and Marvin, 2001: 113).

In the context of development theory, these 'secessionary' infrastructures physically by-pass sectors of cities unable to afford the necessary cabling, pipe-laying, or streetscaping that underpins service provision. Cities such as Manila, Lagos or Mumbai are thus increasingly characterised by a two-speed mode of urbanisation. There is a social geography to all of this, of course. Graham and Marvin highlight the fact that the splintering metropolis has − due to the growth of privatised utility companies − been driven by market rationalities, rather than notions of territorial equity. This idea, where water pipes bypass poor communities in developing countries, or peripheral housing estates lack internet cabling, undermines attempts to provide universal access to services. As such:

> Instead of standardised infrastructure networks operating more or less homogeneously to 'bind' a city, there is a set of so-called 'tunnel effects'. These are caused by the uneven 'warping' of time and space barriers by the advanced infrastructure networks, targeted on valued parts of the metropolis and drawing them into intense interaction with each other. (Graham and Marvin, 2001: 201)

Thus, infrastructure provision is not something that can be taken for granted and its provision and maintenance are closely linked to social inequalities in cities. This has been conceptualised as 'glocal bypass' (Graham and Marvin, 2001), where user-pays access networks allow the linking of highly productive economic spaces while ignoring those sections of the city that are not so productive.

## Cyborg urbanism

Having accepted the significance of how infrastructure dictates the success or otherwise of city life, and rolls out urban speed and service quality into rural areas of national territories, there remains the question of how individuals are 'plugged' into the webs and networks of these services. Gandy (2005) has charted out a series of ways in which this can be achieved. Drawing on Donna Haraway's (2001) theorisations of the 'cyborg', Gandy argues that 'If we understand the cyborg to be a cybernetic creation, a hybrid of machine and organism, then urban infrastructures can be conceptualised as a series of interconnecting life-support systems' (2005: 28). These possibilities of enhancing human capability are taken to their extreme, perhaps, in 'the technologically enhanced soldier of the twenty-first century peering around the corner of buildings in defence of prosperous nations and their corporate sponsors' (p. 32). At a simpler level, many aspects of everyday life, from pace-makers through to interaction with computers can be considered as an extended human (see also Michael, 2006). In this context, cyborg urbanisation emerges 'as a corrective to those perspectives which seek to privilege the digital or virtual realm over material spaces' (p. 40).

A key issue is thus the visibility of such infrastructures. Many infrastructures are often visually excluded from the eyes of the consumer. Hidden from view, buried beneath the ground and with an increasingly a-geographical management regime (London's water supply is currently owned by the Sydney-based Macquarie Bank, for example), the reliance on nature as a resource for the domestic is ignored. As Kaika (2005) demonstrates, the

> function of the modern home as safe and autonomous is predicated not only upon the entry of 'good' nature, but also upon the ideological and visual exclusion of 'bad' nature ... The presence of good water inside the house is based on the existence of a set of networks of, and connections to, both things (dams, reservoirs, pipes) and social power relations

(struggles over the allocation of water, over policies of pricing and privatization) that exist outside the domestic sphere. (p. 64–65)

To conclude, it is important to think through the nature of infrastructure as being taken for granted, but absolutely fundamental in the constitution of modern urban life. With its ability to zip the nation together, to shrink distance, or knit together cities on different continents within the embrace of air networks or telephone and internet connections, it is instructive to think through how the scalar configuration of infrastructure networks builds territorial cohesion. From the grand spatial narratives of empire or nation-building, through to the subtle – though dramatic – modifications to human capabilities offered by networked technologies, infrastructure offers a key to understanding the urbanisation of cities and their wider territories.

For Keeling (2007: 223), this poses challenging research questions for geographers:

How and where do global airline services contribute to the buildup of greenhouse gases in the atmosphere and what are the long-term implications of significant air-traffic growth, especially from rapidly changing economies like China and India? What challenges face global environmental systems as emerging economies adopt North American-style transportation policies and attitudes towards accessibility and mobility? ... New analytical techniques, redefined spatial relationships, enhanced transport capabilities, and infrastructure for a global economy all argue not only for fresh approaches to traditional problems ... but also for innovative ways of conceptualizing and analyzing transportation for the twenty-first century.

Just as the process of moving is important in underpinning the flows travellers – either as tourists or migrant workers, either temporarily moving or going from one sedentary life to another – that can be captured within the frame of **transnational urbanism**, so a study of the systems and processes of embedding infrastructure is a central theme of urban geography.

## KEY POINTS

- Infrastructure has a huge role to play in the functioning and feel of cities, from sewerage systems to power plants, as well as underpinning the connectivity of cities which is often taken for granted by globalisation theorists.

- While seen for much of the twentieth century as a public good, oriented toward universal access, many new developments are based upon a 'user pays' ideology, allowing premium services for those who can afford to pay more. Access to such infrastructure is unevenly divided within cities, with the situation even more extreme in underdeveloped countries, which have often suffered from inappropriate and destructive megaproject development undertaken by Western firms.
- The successful functioning of infrastructure allows individuals to be seen as 'cyborgs'. However, system breakdown threatens the taken-for-granted comforts of everyday life in prosperous cities and reveals the human dependency on natural resources and scientific expert systems.

## FURTHER READING

As noted in the text, Graham and Marvin's (2001) *Splintering Urbanism* is a lively, empirically rich discussion of the reshaping of infrastructure within metropolitan space economies. The legal anthropologist Eve Darian-Smith's (1999) book *Bridging Divides: The Channel Tunnel and English Legal Identity in the New Europe* provides a fascinating discussion of the ramifications of a high-speed rail and tunnel link between England and France, highlighting its transformative impact on the places passed in between. Keith Revell's (2003) *Building Gotham* provides a fine social history of the interplay of private wealth and public policy in the early decades of twentieth-century New York.

*DMcN*

# 2.4 ARCHITECTURE

The study of buildings and their social meanings has been a popular academic pastime. Architectural history, in particular, has a long history within universities and usually seeks to explain the motivations of particular architectural movements and building types, through a careful consideration of their formal properties (their scale, materials, use of space and light, etc.). However, many (though by no means all) of these studies are concerned with the building alone, in isolation from its location in the urban landscape. Geographers are, by the nature of the discipline, required to consider both the building itself but also its implication in wider social networks involved in its production (in terms of the power and wealth of the clients that order its construction) and also its use, bearing in mind that diverse groups use buildings (from social housing tenants to private housing owners in gated communities, to office workers in sky-scrapers to the audiences that visit opera houses or sports stadia).

There is a long tradition of landscape geographers who have sought to explain and account for representations of material landscapes and the social struggles which take place over which representations gain widespread public exposure. J.B. Jackson, a geographer at University of California, Berkeley, played a significant role in documenting the non-metropolitan landscapes of the United States in books such as *Discovering the Vernacular Landscape* (1984) and *A Sense of Place, A Sense of Time* (1994). More recently, however, the growing interest in Marxian geographies began to orient attention away from the specific material qualities of the building or landscape itself, towards an understanding of the building and development process within capitalist economies. Such a perspective – as summarised by Knox (1987) – sought to account for the role of labour in the building construction process, the specific social classes that had the resources to develop and reinvest profits from large-scale building, and the importance of experts such as architects in maximising the profitability of building sites. Most importantly, such approaches moved away from concentration on the visual and material, to a consideration of the apparently abstract processes that allowed for the production and redevelopment of the built environment.

As Jane M. Jacobs summarises below, recent geographical work on architecture and building has often used the building itself as a means of reflecting on the social conflicts and power plays that make possible their construction (and, by extension, their destruction):

> The scholarship on buildings and building events has taken many forms within cultural geography. The theoretically and methodologically dis-parate fields of 'geography of settlement', 'urban morphology', 'urban semiotics', and 'cultural politics of the built environment' all scrutinize the built form in one way or another. The earlier versions of such scholarship privileged the materiality of the building, often requiring it to operate forensically as evidence of wider, more abstract processes or morpholog-ical conditions. In more recent geographies, the technical and formal qualities of buildings served as a faint skeletal infrastructure for studies more concerned with meaning and the politics of representation. (Jacobs, 2006: 2)

The following discussion seeks to summarise some of the key interventions made within geography which have brought together the visual and material qualities of the built landscape and the social processes that underpin its construction.

# Landscapes of power

Around the late 1980s and early 1990s, a particular moment in urban geography and sociology was reached where the rumblings of a 'sea-change' (to use David Harvey's wording) became evident in the British and American economies. *The Condition of Postmodernity* (Harvey, 1989b) uses architecture as a central motif in what Harvey saw as a shift from modernism to postmodernism in cultural production, closely tied to structural shifts within the world economy. Scholars began to consider the social and political production of urban space and to analyse the many forces that structure the built environment, from architects to financiers to politicians to property developers (Crilley, 1993; Fainstein, 2001; Knox, 1987; Zukin, 1992). Both Harvey (1994) and Merrifield (1993a, 1993b) offered trenchant analysis of the fictitious financial instruments that founded – and were switched by agents sitting within – some of London's keynote office buildings.

The work of the French social theorist Henri Lefebvre became very popular at this point, given impetus by the appearance of *The Production of Space* (1991). Merrifield (1993: 1281), for example, argued that Canary Wharf and similar buildings in New York and elsewhere acted as '*spatial inscriptions* of social conflict'. Working within Marxism, but with a humanist perspective on cities, Lefebvre 'attempted to demystify capitalist social space, to trace out its inner dynamics, its *generative* moments, in all their various guises and obfuscations. Here, *generative* means "active" and "creative", and creation is, for Lefebvre, an actual productive *process*'. (Merrifield, 2002: 89). Thus, processes of urbanisation that build and tear up the capitalist city are underpinned by what Lefebvre calls a 'representation of space', 'a space envisioned and *conceived* by assorted professionals and technocrats: planners, engineers, developers, architects, urbanists, geographers, and others of a scientific bent' (Merrifield, 2002: 89). Thus, the discourses and visual imagery produced in the production of the built environment suddenly became of interest to Marxist geographers, used to dealing with more economy-driven (economistic) understandings of society. Lefebvre, therefore, was a key player in bridging these traditions.

However, such work was marked by a tendency to 'read off' social relations from built form. More sophisticated accounts began to emerge in response to this: Jacobs located urban change in the City of London within a more historicised account of the changing spatial relations of the UK economy in her case study of Bank Junction, a major planning struggle against a modernist office building:

> The City of London of the 1980s was both a postimperial city and a postmodern(ising) city; it was a city in transition and change. In this prolonged redevelopment saga, Bank Junction (present and proposed) was invested with meanings by a range of interest groups. The discourses generated by the planning controversy were not simply about the form and the function of this section of the City of London, but also about the renegotiation of the very identity and status of the City in relation to the rest of the world. (1994b: 751)

This introduces an important point: that *debates* over proposals that certain buildings be demolished, altered, or constructed, can reveal some of the deepest-seated identity claims of various cultural groups.

The theoretical structure of the capitalist space economy had been an important theme of Harvey's earlier work, set out at length in *The Limits to Capital* (1982), but he underpinned this with an extensive survey of urbanisation in nineteenth century Paris, recently updated and reprinted as *Paris: Capital of Modernity* (Harvey, 2003). This included an essay on the Basilica of the Sacre Coeur, which – despite earlier attempts to stand as a monument to workers killed in the social rebellion of 1870 (the Commune) – was adopted and resignified by conservative social elites. This is a favourite essay of many readers of Harvey and it shows the significance of monuments and commemoration in the exercise of state power in the built environment. This is closely tied to nation and empire-building, a process visible in Lisbon (Power and Sidaway, 2005), Rome (Atkinson and Cosgrove, 1998) and numerous imperial centres (Driver and Gilbert, 1999). What these works by geographers share – which perhaps differentiates them from the more formal analysis favoured by architectural historians – is an interest in performance. As Edensor has shown, this is a powerful tool in analysing how nations become – in the famous formulation of Benedict Anderson – 'imagined':

81

> by conceiving of symbolic sites as stages, we can explore where identity is dramatized, broadcast, shared and reproduced, how these spaces are shaped to permit particular performances, and how contesting performances orient around both spectacular and everyday sites. (1983: 69)

These are important, because the stolid monoliths of statues, museums, parliament buildings and the wide open spaces of ceremonial squares and parade grounds are animated and choreographed in particular ways: 'Across the world, independence day celebrations, presidential inaugurations, flag-raisings, anthem singing, religious occasions, funerals of important figures, military parades, and "archaic customs" tend to follow the same format year upon year, inscribing history on space' (Edensor, 2002: 73). So it is important to avoid assuming that such landscapes are not powerful in themselves, by mere virtue of their height or design, but rather that social actors perform that power.

## Relational geographies of architecture

Expressive architectural design has taken on an important commercial logic in recent years. With cities riven by deindustrialisation and governments unwilling or unable to raise taxes to bail them out, urban managers such as mayors, chief executives or head planners have come to treat the city as a corporation seeking to gain a market 'niche' in competition with other entrepreneurial city rivals (see **urban politics**). Nonetheless, the rise of the iconic building – captured in such singular statements as Frank Lloyd Wright's Guggenheim in Manhattan, Utzon's Sydney Opera House, or Gehry's Bilbao Guggenheim – can be interpreted as underlining the expressive power of contemporary architecture. Jencks (2005: 185) defines the hallmarks of an icon as including 'the reduction to a striking image, a prime site, and a riot of visual connotations'. It will also benefit from visibility from different angles and perspectives, as well as providing a metaphorical statement (the sail, the pinecone, the fish). This has brought with it a demand for the services of those architects perceived to be possessive of a strongly identifiable visual design, or 'signature'. Given that images of cities are constantly being produced, distributed and consumed, in magazines and newspapers, on television programmes, on postcards or souvenirs, such buildings can help to identify and 'mentally map' places within everyday knowledges.

The authors of these iconic buildings – Hadid, Piano, Meier, Koolhaas, Gehry, Foster, Calatrava, among others – are sought out to rebrand, reposition or otherwise publicise the cities of advanced capitalism. They have been identified as part of a 'star system' of architects with a recognisable 'signature' skyscraper style that appear

**Figure 2.4.1** 'Burj al-Arab, Dubai: the use of iconic designs to brand places is a growing trend in architecture and urban development'

in projects around the world (McNeill, 2008a). A very significant theoretical development of this idea is provided by Kris Olds, who charts 'an emerging (and amazingly small) epistemic community of developers, architects, planners and academics who draw upon each other (or each other's work) in planning and building UMPs [urban mega-projects] throughout the world' (Olds, 2001: 27). Drawing on case studies of Shanghai and Vancouver, Olds illustrates how design knowledge is transmitted and circulated via global consultancy firms, advertised via satellite broadcasting and validated by 'foreign' experts.

The growing trend in understanding urbanisation as a relational process requires a view of specific, material spaces as being switching points or containers of people and technologies that are inter-connected with other similar spaces many miles distant. As with airports and communication towers, skyscrapers are obvious candidates for channelling globalised flows, whether metaphorical or material (McNeill, 2005). Yet this 'global' history of the skyscraper conceals a range of complex relational geographies, from a conventional locational geography (the distribution of tall buildings in the world's cities), to a mobile range of visual codes and corporeal movements, to debates over the nature of transnationalism, global–local relations and ideals of a 'universal' building style popularised by the modernist movement (Jacobs, 2006).

# Global and local architectural form: the skyscraper

Skyscrapers act as central symbolic structures, as landmarks, which orient citizens. They come to symbolise the changing histories of cities, just as cathedrals and castles dominated cities of past times (and continue to – see Sidorov, 2000). Given the complexity and contingencies of global flows, it is ironic that many of the metaphors, adjectives and tropes used to represent and talk about the skyscraper emphasise fixity, solidity, rootedness and permanence.

Within such buildings, the rather crude 'global–local' construct recurs frequently as practitioners and critics alike seek to interpret the uneasy interplay of standardised building production systems, centuries of indigenous design history and relatively distinct modes of living and working. This presents particular challenges in the developmental states of Southeast Asia that have explicitly adopted skyscrapers and infrastructure projects as symbols of national modernisation. For example, Kuala Lumpur's Petronas Towers were designed by an international architect (Cesar Pelli) with Islamic motifs incorporated into the façade and floor-plans, an attempt to fuse standardised western production methods with a locally-sensitive design vocabulary. Bunnell (1999, 2004a, 2004b) argues that the towers were explicitly used to advertise a distinctive Malaysian modernity through easily-quotable iconic architecture that would feature in adverts, in-flight magazines, postcards and even Hollywood feature films.

So, while skyscraper technology is predominantly exported by western firms (Cody, 2003), the design process may be significantly influenced by context-specific factors, be they climatic, aesthetic or cultural. For example, an obeisance to *feng shui* guides many Southeast Asian skyscraper designs, as exhibited in Norman Foster's HSBC building and the competing Bank of China design by I.M. Pei, both in Hong Kong. Similarly, tall building design in earthquake-prone areas is a key factor in countries such as Japan and Taiwan, and many of the 'critical regionalist' responses offered in Southeast Asia are centred around climatic concerns, such as Ken Yeang's 'bioclimatic skyscraper', which gives us a range of new forms driven by the need to capture winds and provide shading.

The materiality of the skyscraper – its height, its form, its massing, its footplate, infrastructure and neighbourhood – endows it with a special place within urban territories. Socially, it opens up numerous questions about the nature of the transnational knowledge flows and how barely visible material transactions are *housed*. Collectively perceived as a skyline, it is able to horizontally define cities in a convenient representational frame, exploited by film-makers, politicians and architects alike. Its sheer verticality raises questions about urban futures, the art and work of living high, but also demands an attention to roots and the invisible cities of service areas and underground transport (McNeill, 2005).

# Everyday landscapes

Commercial or state-sponsored monuments are visually arresting and can threaten to occlude a discussion of the private, or domestic, particularly in terms of home and housing. In this context, geographers are using a wide repertoire of theoretical insights to interpret the attachments, aesthetics, materiality and representations of the private sphere. Blunt and Dowling identify what they term a 'critical geography of home' where they argue for a '*spatialized* understanding of home, one that appreciates home as a place and also as a spatial imaginary that travels across space and is connected to particular sites [and for] a politicized understanding of home, one alert to the processes of oppression and resistance embedded in ideas and processes of home' (2005: 21–2). The examples and case studies that they employ – ideals of frontier home-making in American nation-building, the spread of domestic technologies, the significance of home to diasporic communities – are illustrative of the significance of such apparently private spaces in the constitution of society. This brings methodological challenges too, and scholars have developed the methodologies of 'house biographies' that reconstruct the personalised meanings of homes, and have called for a recognition of the polyvocal nature of housing (Llewellyn, 2004).

However, by contrast with the analysis of a static building (by choosing a single building and unpeeling the layered narratives of its construction and reception), there is a growing interest in explaining and documenting the ways that spaces and landscapes are constituted in motion. For example, Merriman's (2006) explorations of the role of landscape architects in the constitution of the post-war British motorway system are illustrative of how the hugely disruptive technologies of high-speed roads were integrated into the English countryside, reconciling the visions of relatively unchanging rurality with motorways that were 'designed around the movements and embodied vision of the high-speed motorist' (p. 85). The apparently mundane structures which abut the motorways – service stations, bridges, and police points – were all debated with the same seriousness as the visual codes of central cities: 'landscape architects stressed that it was the task of engineers and landscape designers to understand the driver's mobile gaze in order to design roads which were *not striking* and would *not distract* drivers attention from the events unfolding on the road' (p. 95).

Adapting a 'mundane' approach to architecture allows for a deeper understanding of individual attachments to home, such as interior design (Leslie and Reimer, 2003) and gardens (Power, 2005). It involves a deeper understanding of the motivations of inner-urban dwellers (Allen, 2007) and property developers. For example, Fincher (2007) traces the assumptions of developers of high-rise apartments in central Melbourne, revealing that despite a veneer of innovation in matching housing products with newly formed groups of 'empty nesters', most developers retain conservative assumptions of the household types that will buy or tenant their properties. With a changing workplace, there is also a trend towards increased adaptability in housing design, with the potential for home offices now being actively built into developers' calculations. Crabtree (2006) explores the possibilities of eco-city housing forms, organised around feminist principles of care and the integration of workspaces into the domestic arena.

A further perspective seeks to focus on the 'intricate processes of adaptation and possession that take place in homes', processes that are not covered by only focusing on the production or consumption of built spaces (Llewellyn, 2004: 229). By placing an experiential dimension at the heart of the built space, such theorists are making a valuable contribution to the study of how completed buildings are used. In a similar vein, Lees' (2001) account of the varied uses of Vancouver library, or Llewellyn's (2004) reconstruction of how residents of an ideal-type modernist housing scheme in London understood their experience of living there, are pointers towards the gap between technical products and everyday spaces (see also Jacobs et al., 2007). For Llewellyn (2004: 246), 'The academic worth does not end once the architects have finished and the photographs have been taken for the journals. In many ways that is the starting point for a critical geography of architecture'.

## KEY POINTS

- A tension between materiality and representation has characterised how geographers have thought about buildings for some time, but increasingly there are moves to synthesise and reimagine the connections.
- Geographers have been interested in explaining the mobile and relational production of architecture, and the technological systems that underpin its design, construction and maintenance.
- Perspectives from landscape geography have shifted from a fixed accounting of visual artefacts to a deeper, archive-based interpretation of the social, economic and political processes that shape the built environment.

87

## FURTHER READING

Deyan Sudjic's (2005) *The Edifice Complex: How the Rich and Powerful Shape the World* is an accessible, if sometimes anecdotal, account of the relationship of power elites to architects and buildings. *Imperial Cities* (Driver and Gilbert, 1999) contains a selection of papers that illustrate the embeddedness of imperial rule in architectural and the built environment in a range of cities. Tim Bunnell's (2004b) paper 'Re-viewing the *Entrapment* controversy' is a fine example of the relationship between the materiality of built form and the representational power of Hollywood.

DMcN

# 3 Envisioning and Experience

# 3.1 DIAGRAM

Many high school or second level students first encounter urban geography through particular kinds of line drawings or diagrams. Generations of students – including some of the contributors to this book – learned about the spatial pattern of cities through exposure to three diagrammatic representations of the city used by Chauncy Harris and Edward Ullman (1945) in their now classic article on concentric-zone, sector and multiple nuclei models of urban growth (see figure 3.1.1). Even if Harris and Ullman's models have been critiqued and revised, the technique of diagrammatic representation upon which they depend remains central to the teaching and learning of urban geography: most introductory textbooks about urban geography will contain a range of similar renderings of urban space (see, for instance, Knox and Pinch, 2006). Such books also contain photographs, words and other forms of representation: yet diagrams (and maps) seem to have played a particularly influential role in the emergence and reproduction of urban geography. While diagrams are of particular importance within urban geography, their status reflects two characteristics of geographical knowledge more broadly: first, the centrality of visual representation; and second, the cartographic imperative underpinning the claims of the discipline to produce truthful, accurate knowledge. While such claims have been contested in recent decades, the diagram remains important within urban geography, albeit in ways that Harris and Ullman might not necessarily recognise.

90

## Defining diagrams

What, then, is a diagram? At its simplest, a diagram is a schematic presentation of a certain set of processes in the world. One way to understand a diagram therefore is as a representation that stands in for – and takes the place of – something else, to which it then refers. Of course the distinction between different kinds of representation is not always clear, as for instance in the case of maps and diagrams, both of which are sets of lines that can be understood as representations of the world with varying degrees of accuracy and simplification. Furthermore, there are many types of diagrammatic representation.

Diagram

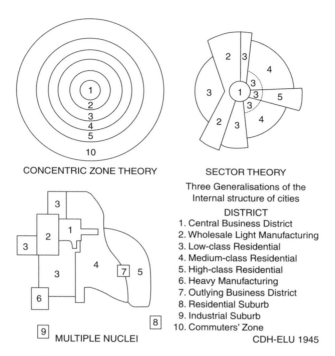

**Figure 3.1.1** Generalisations of internal structure of cities. The concentric-zone theory is a generalisation for all cities. The arrangement of the sectors in the sector theory varies from city to city. The diagrams for multiple nuclei represents one possible pattern among innumerable variations.

For instance, the following are only a few of the types listed and discussed by a textbook from the early 1970s: ray-diagrams, ranking diagrams, stage diagrams, time–space diagrams, regression diagrams, scatter diagrams, and spatial association diagrams (Monkhouse and Wilkinson, 1973).

Rather than look for a simplified definition of what a diagram is we might be better served by defining it in terms of what it does and, more specifically, what it allows urban geographers to do. First, and most obviously, diagrams offer an important technique of abstraction. They provide a way of extracting from the dynamic flux of urban life a schematic presentation of certain elements of that life in the form of points, lines or sections. Second, and relatedly, diagrammatic representations are techniques of analytical simplification and selectivity: they allow for certain urban processes to be isolated, even if only partially and temporarily. Then, and third, diagrams allow for the possibility of generalised comparison between or within different urban spaces or urban processes. So Harris and Ullman's diagrams can, for instance, be used as the basis for making comparisons between the spatial development and growth of different cities (see Lichtenberger, 1997). Fourth, diagrams have a projective quality: they allow for the imaginative or calculative anticipation of what might happen in the future based upon identifiable trends or processes in the past. Finally, diagrams (with maps) are very much part of the public face of a social scientific discipline such as geography. They convey information in ways that go beyond some of the limitations of text.

## Diagrams and urban geography

Given their usefulness it is hardly surprising therefore that diagrammatic representations of cities have become central to how urban geography is understood. While such use of diagrams can be traced through the influence of a range of geographers and non-geographers (most notably the Chicago School), they have been associated particularly closely with efforts to transform geography (and within it urban geography) into a kind of spatial science. Such efforts were expressed most systematically during the quantitative revolution of the 1960s: here the emphasis was on the development of quantitative models of geographic processes and patterns, the results of which could be more accurately represented in graphic or diagrammatic form. The apparent objectivity and neutrality of such representations underpinned the claims of spatial science: certainly, diagrammatic representations seemed less influenced by subjective factors than written text. As such, the diagram could be employed as part of a system of knowledge production that aimed towards the production of generalisations and predictions about urban space.

While the heady days of the quantitative revolution have passed, diagrammatic representations of cities remain influential and in use, albeit in a more dynamic form. Computer simulation and modelling have enhanced the complexity and accuracy of attempts to represent urban processes and spaces (see, for instance, the pages of *Environment and Planning B: Planning and Design*). Yet while the technical capacity of urban geographers to represent cities has undoubtedly improved, the very role that such representations play in the process of understanding cities has come under increased critical scrutiny. This scrutiny has emerged in part from a growing appreciation of the possibilities and problems of various representational renderings of urban space. One of the most important critiques levelled at the kinds of diagrammatic techniques employed by the quantitative turn in the 1950s and 1960s is that they reduced human experience to a kind of idealised abstraction; everyday attachments to place, themselves the product of routinised practice, tended to be excluded from models of urban space and process.

The case of time-geography offers an influential and instructive example of an attempt to negotiate this problem: it can be read as an effort to rework diagrammatic representation of spatial patterns and processes such that this representation included the experiential dimensions of everyday life within an analysis of wider spatio-temporal structures. Emerging in the early 1970s, time-geography is based upon the work of Swedish geographer Torsten Hagerstrand (1970, 1982), developed further by geographers such as Nigel Thrift and Allen Pred. One of the key elements of time-geography is the use of a series of diagrammatic techniques for tracing systematically the spatio-temporal paths of everyday life in order to reveal the dynamic pathways of the individual in relation to a range of networks of differing scales and intensities. Yet, by the end of the 1980s, time-geography had fallen out of disciplinary favour, in part because of a number of criticisms levelled at the manner in which it tried to represent everyday urban life. Perhaps most importantly, feminist geographers such as Gillian Rose argued that time-geography remained wedded to a mode of abstraction that failed to account for the multiple ways in which urban space is inhabited: nor could it accommodate the different kinds of bodies that negotiated urban space. Indeed, for Rose (1993: 31), 'time-geography tries to ignore the body' by reducing it to a pathway of movement.

While the veracity of this critique is debatable (see Latham, 2003; Thrift, 2005c), it serves nevertheless to exemplify a broader disciplinary engagement by geographers with the limits and politics of representation. Such engagement emerges from a critique of spatial science and the confluence of theoretical developments that brought together insights from cultural studies, Marxist thinking and humanistic understandings of space and place. This critique of representation has a number of elements, the sophistication of which has evolved over time. First, there has been a sustained attempt to explore the role that representations of the city have played in shaping the material production and imaginative understanding of urban space. Second, geographers have challenged claims that representations of urban space are ever neutral: they are instead active participants in relations of power. Third, as part of the wider emergence of non-representational theory, key figures within the discipline have argued that much of what happens in cities does not rely upon representation at all (see Amin and Thrift, 2002). This last point is of particular importance because its implications are often misunderstood. To claim that much of urban life does not involve representations does not mean that diagrams are unimportant: rather, it means thinking of diagrams less as static snapshot-like images that freeze process, but as presentational techniques that perform (and do) various kinds of work.

# Reanimating diagrams

This understanding of diagrams – as participants in process – has a number of important precursors. The first is philosophical and can be traced through the writings of thinkers such as Michel Foucault (1926–1984) and Gilles Deleuze (1925–1995). Both push the meaning and power of the diagram beyond its narrowly representational focus. Foucault is interesting here because he understands the diagram as a kind of political technology – a device for arranging and distributing things in space and time with powerful effects. Taking this further, Deleuze (1999) argues that the diagram is a kind of 'abstract machine', generative of certain predictable stabilities and continuities in forms of conduct. Why 'abstract'? Because the diagram is not necessarily locatable in one social institution or indeed in any one place: nor indeed is it something controlled by any single individual. Deleuze also uses the term 'machine' in a particular sense: it is not a physical object but a set of connections between different processes. Such connections might not be actually visible: for instance, we might think of desire as a kind of machine that generates connections between bodies. And, as the case of desire demonstrates, the invisibility of machinic relations does not render them any less powerful. For Deleuze, following Foucault, the diagram is powerful because it has the capacity to form and order connective relations and across societies.

A second influence on efforts to rethink the concept of diagram in urban geography is the set of ideas known as actor-network theory, often associated with figures such as Michel Callon, Bruno Latour and John Law. In the work of such thinkers the diagram is understood in a more obviously literal way than in Foucault or Deleuze: diagrams are active participants in the production and maintenance of heterogeneous networks. Diagrams hold things together while also allowing them to circulate across space and time. When we look at a diagram we know the things or processes to which it refers are not literally there: yet at the same time the lines of which the diagram consists seem to have transferred something of the qualities of whatever that particular thing or process is. So while the thinkers like Latour are interested in how actual diagrams work, they also want us to think about how this work involves a sequence of transformative and traceable operations in which the diagram 'replaces without replacing anything [and] summarizes without being able to substitute completely for what has been gathered'. The diagram for Latour is therefore a strange 'alignment operator, truthful only on condition that it allows for passage between what precedes it and what follows it' (1999: 67).

These ideas about the diagram have been taken up by geographers such as Amin and Thrift (2002) in an attempt to rethink the relation between power and the city (see also Latham and McCormack, 2004). For such geographers the concept of the diagram is appealing in the sense that it does not reduce power to the terms of an ideological or material structure through which order is imposed from above: rather, diagrams are productive and therefore powerful insofar as they infiltrate and order a range of practices at the micro scale. One way to make this a little more digestible is to think of the role that engineering or architectural diagrams play in the construction of buildings and roads (see Robertson, 2007) or, on a larger scale, the planning of urban areas. These diagrams are not simply outline representations of objects: what they do is allow a set of objects and agencies to go through a series of transformations such that a building or set of buildings, roads, etc. emerges. These engineering diagrams also draw together and depend upon a range of other diagrams – those concerned with sewerage, electricity, water, transportation, etc. The diagram in this sense acts as a mode of ordering that provides for movement between different material and symbolic registers.

At the same time, because such diagrams are projective they also provide a technique for inhabiting possible futures. This is particularly obvious in relation to architecture. Within architecture (and design), the diagram can be understood to have a purely functional role: it can be read as a plan or as a simplified schematic representation produced for a client. But recent writings by architects and architectural critics points to the diagram as something with a more utopian potential: it provides for the imaginative construction of spaces that might rework relations between people and their environment. We might observe that such claims have always been made by and of architects, particularly those (such as Le Corbusier) interested in effecting social transformation through architectural intervention on a large urban scale. Drawing upon thinkers such as Deleuze, more recent architectural writing has focused on how the diagram becomes generative of a range of possible futures, providing an unfolding context for action (see Vidler, 2001) through a complex set of negotiations between architectural and other practices (see Allen, 1999).

## Making more of diagrams

So the work of thinkers such as Foucault, Deleuze and Latour provides urban geographers with ways of thinking about how diagrams facilitate all kinds of spatial transformations. They also invite us to think of diagrams as productive – as things that do work, that transform, that perform. Geographers are beginning to take up this invitation in at least three ways. First, by employing the diagram as a conceptual device through which to think about how diverse agencies and actors can be drawn into an account of distinctive urban ecologies. As Hinchliffe et al. (2005), in a discussion of various urban wild spaces argue, the activity of diagramming provides a way of making such spaces cohere, if only loosely and temporarily, as real matters of debate and interest. A second way in which geographers have begun to take up the invitation to experiment with the diagram as a conceptual device is in relation to spaces of bodily performance and practice. Here the diagram becomes a way of thinking through the articulation between bodily and architectural space (McCormack, 2005: see also Imrie, 2003).

Third, geographers have begun to revisit various diagrammatic techniques in order to reaffirm the value of creativity and inventiveness as necessary components of thinking about and through the dynamics of urban space. Potential sources of such inventiveness include, for instance, the tradition of cartographic subversion and experiment associated with situationist thinkers (see Sadler, 1998; Pinder, 2005). Additionally, once dismissed because of their abstract quality, the diagrams of time-geography might now be seen in a more generous light

# Diagram

**PAUL: A WEDNESDAY IN JUNE, 1999**

"I read until I'm off the Planet." JvB

2.00

"I used to go out with a guy who smoked in his car..." LL

"Stop waving your wand, Lyn." PCB

23.00

"I take the cigar out of my mouth before I kiss a woman." RB

22.00

ARMADILLO HAD OPENED JUST A FEW WEEKS BEFORE DESIGNED BY PATRICK STEEL, THE PLACE IS AN EXPENSIVE FUSION OF TEX-MEX AND MODERNISM ...

"Armadillo was kinda a by-product of a cocktail party we had at the house. The conversation was that we should have a dinner party. I was going to cook. Then I went to Armadillo and I thought 'Stuff the cooking! Why don't we just do it here?'"

21.00

Drives to Armadillo on K'Rd, just off P' Rd, with Linda his girlfriend for a meal and drinks he's organised with a bunch other late-30 and 40 somethings ...

20.00

"All SOLO PARENTS WITH NO KIDS. Aspwnks."

THE CAFES ALONG PONSONBY ROAD ARE IMPORTANT PLACES FOR MEETING CLIENTS AND JUST SEEING WHAT IS GOING ON...

19.00

18.00

"You know, it might look like you're mucking around but you're not. You're laying your self open to everything and anything that's going on around you, and that's where the people are! My business is about being where the people are...."

17.00

16.00

Drives to meeting with a client in Ponsonby. Returns to work after 30 minutes or so...

PAUL'S A REAL ESTATE AGENT. THIS MEANS HIS HOURS ARE FLEXIBLE, AS—TO A DEGREE—IS HIS PLACE OF WORK... NONETHELESS, MAKING MONEY IS ABOUT BEING ORGANISED; MAKING THE PHONE CALLS, BORING TEDIOUS STUFF...

15.00

"Personality, the supreme realisation of the innate individuality is an act of the greatest courage" Carl Jung

**KEY**

Driving

Walking

Activity

14.00

13.00

She came out of Toffee designs. Brown paper bags to look like they came from Paris. She was immaculately dressed, and swung her hips as she crossed Jervois Rd."

Finally gets to work

12.00

Drives the short distance to work. But makes what turns out to be over an hour detour chatting to a friend at Gannet Rock. He's feeling hung-over and lethargic...

11.00

Home (St Mary's)

WorkPonsonby (Ponsonby Road)Road

Time

97

**Figure 3.1.2**   Latham (2003a) provides an example of how an individual's daily routines can be captured in a diagram, mixing photos and text drawn from diary entries.

insofar as they seem to have anticipated some of the more recent interest in the diagram emerging through encounters with thinkers such as Deleuze and Latour (see Thrift, 2005c). With some creative reworking, they can be made to incorporate a range of experiential detail about everyday urban life (Latham, 2003). (see figure 3.1.2) And technological developments since the first wave of time-geography provide a much more flexible and potentially more empowering use of diagrammatic techniques (see for instance Kwan, 2002). Used in this way, the diagram becomes a kind of cartographic political technology, through which a range of potential urban futures might be experimented with.

Perhaps most importantly, the inventive dimensions of techniques like time-geography remind us that the diagram does not designate a solely representational technique. The diagram may be presentational: it is also, however, aesthetic, affective and kinaesthetic, produced by and generative of particular movements of thought, hand or pen. Such claims might well be read as hyperbolic: applicable only to the more avant-garde kinds of diagrams encountered in architectural or artistic practices. Yet, and perhaps with a little more modesty, they can also be considered to apply to the kinds of line diagrams contained in hundreds of introductory textbooks on urban geography.

98

## KEY POINTS

- As a schematic presentation of processes, the diagram has and continues to be central to the production and reproduction of urban geographical knowledge.
- While the diagram has traditionally been understood as a representational technique, more recent work within human geography has focused on diagrams as performative devices – devices that perform certain kinds of work.
- In turn, drawing upon such work, it is possible to reaffirm the creative element of the diagram when thinking about urban space.

## FURTHER READING

Harris and Ullman's (1945) paper on *The Nature of Cities* is a classic example of how urban patterns and processes can be rendered diagrammatically. The work of Torsten Hägerstrand (1982) provides the most obvious point of departure for thinking about the diagrammatic concerns of time-geography. Alan Latham's (2003) *Research, Performance, and Doing Human Geography: Some reflections on the diary-photography, diary-interview method* provides an example of how urban geographers are revisiting and reworking diagrammatic techniques in the wake of more performative approaches to methodology. The discussion of 'urban wild things' by Steve Hinchliffe, Matthew Kearns, Monica Degen and Sarah Whatmore (2005) illustrates how a philosophically informed understanding of the diagram supports a rethinking of the politics of urban nature.

*DMcC*

# 3.2 PHOTOGRAPHY

How do we come to know cities without having physically experienced them? Visual images of cities can travel to us through television, the cinema and photographs, which are then distributed through various forms, such as cinema multiplexes, DVDs, the internet, newspapers, magazines, postcards and websites. However, it is perhaps not images of cities as a whole, but parts of them – such as buildings, streets and bridges – that are most widely known and come to represent the city as a whole. Select components become familiar and 'iconic' images of the city are fed into the popular imagination of viewers. For example, the Opera House can be instantly mentally 'mapped' to Sydney, the Eiffel Tower to Paris, Big Ben to London and the Golden Gate Bridge to San Francisco. But, of course, there are not only visual images *of* cities, but also visual images *in* cities, for example, billboards, posters, TV screens and graffiti. (For further discussion on this, see the **media** entry.)

Visuality and an interest in visual culture more generally, has been related to the apparent dominance of visual images over textual or verbal modes of communication. This entry doesn't intend to enter this debate, but does seek to focus on the significance of photographic imagery and practice within contemporary urban life. To consider this

**100**

**Figure 3.2**    Fashion retail adverts on the Tokyo skyline

further, a useful categorisation of the spatial sites of visual imagery is provided by Gillian Rose in her book *Visual Methodologies* (2007). She differentiates between the site of the *production* of an image (which can range from a film studio to a location set to a home computer), the site of the image itself (its formal compositional properties such as colour, size of its subject matter, its framing, etc.) and the site where the audience sees the image (understood as an artefact which moves about in space, such as in a magazine, on a postcard or in an art gallery.) This opens up an important way of considering the social meaning of images, made more important because of the widespread use and distribution of photographic (and cinematic) technologies, which effectively 'globalises vision' (Schwarzer, 2004). Furthermore, photography acts as a distancing mechanism, and with the historical increase in speeds by which cities – and events in cities – can be viewed (and now edited, repositioned and redistributed), the visual landscape becomes replaced by a technological 'zoomscape' (Schwarzer, 2004).

# Photography and the city

To understand the nature of this more fully, we should consider the historical evolution of photography and the city. In the genre of 'street photography', which has evolved in a fascinating direction over the twentieth century, new and innovative framing techniques, combined with technological advances, have transformed how the social life of cities is documented. For example, Paris – a capital of modernity in painting, music and fashion in the early twentieth century – became an important site of early urban photography. Eugene Atget, Henri Cartier-Bresson and Robert Doisneau were influential pioneers of street photography, whose work provides some of the most iconic images of urban life. Working in the first half of the twentieth century, Cartier-Bresson's photographs have come to represent what we could think of as 'original' or 'authentic' images of Paris. Photographs such as *Behind Saint-Lazare station*, Paris, 1932 and *Michel Gabriel, Rue Mouffetard*, 1952, have been used as symbolic representations of Parisian life in the early and mid-part of the century. This might be because Cartier-Bresson's technique consisted of spontaneously responding to 'whatever he found' while walking the streets. Cartier-Bresson's photographs are also typically 'gray and even in tone' (Westerbeck and Meyrowitz, 2001: 154), as he never photographed with a flashbulb and worked exclusively in black and white.

101

This aesthetic is a way in which viewers can first see – and then imagine – the streets of Paris, which adds to the power of these images as both art objects and as anthropological artifacts, mediating 'interactions between people and visual objects' (Rose, 2007: 217). Along with this, the materiality of the photograph must also be taken into consideration: 'how they look and feel, their shape and volume, weight and texture … it also implies placing photographs in particular geographical locations and in their social and cultural contexts.' (Rose, 2007: 219).

So, it is important to consider not only what the photograph represents two dimensionally, but also where and how (and on what type of object) the image is processed and travels. Tracking the *object* that carries the image is a good way of assessing the image's significance (Rose, 2007: 230) For example, Cartier-Bresson's original photographs, taken and developed in Paris, are exhibited in galleries all over the world; postcards featuring images of his photographs might be sent from France to Australia; printed posters could be manufactured in China and sold in America. Postcards are a particularly interesting example of how specific parts or essences or 'myths' of cities are framed and distributed to wider audiences, tending to play on the stereotype of a place, often skewing the interpretation to a particular historic, political or sociological context. They are, of course, consumable images for the tourist, whose travels are supposedly reflected in these mass-produced visual souvenirs. As Waitt and Head (2001: 320) note:

> Carried in and through pictures and languages are social constructions, or naturalised, commonsense views or 'ways of seeing' the world that are more than the sum of the words or imagery. Images of places are never taken as straightforward mirrors of reality. Instead, the meanings of an image are understood as constructed through a range of complex and thoroughly social processes and sites of signification. Places depicted in postcards are no exception and are also more about myths than about substance. Picture postcards are endowed with symbolic meanings. Postcard imagery is one mechanism by which tourism places are reinvented in the image of particular tourist motivations and desires.

**102** When cities are portrayed in moving images, but combined with fictional storylines as in films and television series, the relationship between representation and reality is even more complicated.

## Screenscapes

> The American city seems to have stepped right out of the movies ... To grasp its secret, you should not, then, begin with the city and move inwards towards the screen; you should begin with the screen and move outwards towards the city. (Baudrillard: 1988: 56)

Geographers have become interested in how to capture a sense of mobility in the urban landscape; or, on the other hand, to represent fixity within the moving image. Prompted by social theorists such as Jean Baudrillard (1988), who has been influential in setting up theoretical frameworks from which we can understand the representation of American cities in film and television, the theorisation of 'cinematic cities' has been an important field of enquiry (see for example, Clarke, 1997; Balshaw and Kennedy, 2000; Sanders, 2001; Shiel and Fitzmaurice, 2001; Barber, 2002). As David Clarke notes, American cities in particular are commonly represented cinematically and televisually, which he suggests contributes to 'a conceptualisation of the cityscape as a *screenscape*' (Clarke, 1997: 1).

The city is often used as a plot device and a backdrop for movies. For example, establishing shots not only indicate the theme, but also work towards providing the identity of the city in which the movie is set, without dialogue. A film only needs an establishing shot – perhaps of a skyline – and a few lines from actors, to help show an audience where it is set. As Pascal Pinck notes:

> Although we may recognise the typical forms of our home metropolis on screen – especially if these forms are visually distinctive as in New York or Seattle – television's architectural cues revise our *idea* of the city: what it looks like, how to live there, and how it may be navigated. Framing, lighting, captioning and graphics provide not a simple representation of urban space, but a hybrid that emerges simultaneously in the picture tube and the street: a televisual city. (Pinck, 2000: 56)

Film can also stereotype and create mythical identities of cities. Mike Davis in *City of Quartz* (1991) and *Ecology of Fear* (1998) uses a framework derived from film noir to set up his political economy of urban restructuring in Los Angeles, using films such as *Chinatown* and *Blade Runner* to characterise the power structures of the city at various times. As Bunnell (2004) shows in his case study of the Hollywood film *Entrapment*, starring Sean Connery and Catherine Zeta-Jones, the use of Malaysia's Petronas Towers (promoted as a symbol of Malaysian modernity by that country's government) were undermined through a cinematic splicing with scenes of slum dwelling, provoking a significant public debate.

This demonstrates the idea of what visual theorists call synecdoche, when an image stands in as a representation of a larger entity. This can be fairly simple, where, for example, an image of the Eiffel Tower provides an instant visual clue that a news broadcast or fashion shoot is located in Paris. Films shot in New York frequently utilise the city's skyline as a backdrop, and other famous architectural landmarks such as the New York Public Library and steaming man holes. Some use landmarks and landscapes as integral sites of the unfolding of the film's storyline. For example, *King Kong* uses the Empire State Building as a pivotal narrative device for its plot. In the film the building becomes more than a prop and it could be suggested that Kong's conquering of the skyscraper symbolises the idea of the ape's control over the metropolis as a whole. In

103

this case, the skyscraper becomes an iconic symbol both of the city's power and its vulnerability (Sanders, 2001: 41).

Another element of the screenscape is the physical presence of image projectors, or screens, within cities. In her book *Ambient Television: Visual Culture and Public Space* (2003), Anna McCarthy argues that television may constitute both public and private spaces. On one hand, television screens in public spaces – such as in airports, waiting rooms, or restaurants – are meant for everyone and, on the other hand, they are obviously private property, as they are often screwed down, or otherwise out of reach, disallowing contact or control over what is being shown. (McCarthy, 2003: 121–2). Unable to be changed, they are a means of allowing the commercial penetration of public space, particularly through the sponsorship of large screens in public parks, or the subtler cross-branded presence of smaller screens in malls and shop windows. This also adds to the complexity of theorising **public space** solely in terms of human access and use; the visual realm within such space is subject to powerful commercial forces, which in turn opens up interesting debates about how such spaces may be interpreted.

104

# Location filming

Location is an important concept in urban geography. However, to suggest that films and television programmes may be made 'on location' open up a number of questions concerning the reality of images. For example, the filming of *Entrapment* adjacent to Kuala Lumpu's iconic Petronas Towers did not have the desired effect that the Malaysian government sought, effectively 'orientalising' the city by cutting in scenes and frames of underdevelopment filmed 150 kilometres away in Malacca (Bunnell, 2004b).

This is a standard trick in the film-making business. Other cities, such as Toronto and Sydney, have also been used as stand-ins for 'American' cities, such as New York, mostly for economic reasons. Australia and Canada often provide much more economic value for making American films. This can be for a number of reasons: the Australian and Canadian dollar has often been weaker than the US dollar, film crews are not as well paid or unionised as they are in the US, and with the money saved on these elements of production, more can be spent on procuring top-name actors and directors, thereby providing a potentially more lucrative outcome.

And so, the city can take on a number of meanings, narratives, genres, and personalities, which gives audiences a sense of what it might be like without having the need to travel to it. Indeed, these representations are often highlighted, via television and magazines, by tourist boards to create a certain type of image of a place, to *entice* people to travel. Television holiday programmes are one such genre:

> In the construction of narratives of place there have been parallels between the scopic, especially the photographic, practices of the tourist and those of the television camera. Broadcasters have used power/knowledge strategies to their own advantage in gazing on and constructing an Other, while often through linguistic or other inadequacy, failing to effect an adequate translation of cultures beyond a recycling of familiar images ... people do not see a place but a succession of sights, authenticated by being reproduced. (Dunn, 2005: 156)

The well-constructed commercialised image of a place can of course be interpreted in contrast to other 'scopic' representations, such as the tourists' own photographs, which often act as highly individualised objects of memory and proof of experience (Crouch et al., 2005; Larsen, 2005).

In a similar vein, the fictional 'televisual city' itself often becomes a tourist destination, further altering individual interpretations of mass mediated representations. Nick Couldry uses the example of the Granada Studios Tour in Manchester, 'the largest, best-known and most established media location in the UK' (Couldry, 2000: 67). Here, the popular British soap opera *Coronation Street* is filmed, and it is the primary reason visitors are drawn to the site, where 'People pay to visit [the] location they have already watched free on television for years; part of the pleasure is not seeing something different, but confirming that the set is the same as something already seen.' (Couldry, 2000: 69). Already a fictional representation of a real location, the Street is re-imagined by the swathes of visitors it incurs on a daily basis:

105

> People take photographs and are photographed at points of interest ... (They) compare the details of the set with their previous image of the Street ... It's enjoyable to pretend, for a moment, that you live on the Street, posing with door knocker in hand or calling up to one of the characters. (Couldry, 2000: 70)

Though it may not exactly resemble the street they see on television (there are no characters to talk to, no proper interiors behind the facades) it is a satisfying representation of it. In this sense, *Coronation*

*Street* becomes an ersatz tourist destination. Rather than there being a divide between fantasy and reality, people are attracted to the set by being able to explore the media artefact, enhancing their status as active media audiences. Most importantly, this exploration can be commemorated by taking their own photographs, which serve as highly personalised souvenirs.

## KEY POINTS

- There is no set way in which to think about still and moving photography and cities. Improving camera technology, trends in popular culture and the ease of distribution of images on a global scale, makes photographic practice a central aspect of contemporary visual culture.
- Photography as a practice can also be understood as the manipulation of images. However, while geographers have tended to deal with real and representational spaces as completely separate entities, they have become interested in their interplay.
- Film and television not only impact cities by the way of their visual representations, but also have an economic and social impact, particularly through tourist practices.

## FURTHER READING

Mitchell Schwarzer's (2004) *Zoomscape: Architecture in Motion and Meaning* moves through the relationship of photographic imagery and travel – on trains, cars and planes. It analyses the evolving technology of the image through photography, film and television, focusing on each theme with specific reference to the moving image. In *Celluloid Skyline: New York and the Movies*, James Sanders (2001) looks at New York City as a backdrop and a featured character in the film industry. He describes directors' fascination and love for the city as a location, and lists the many and varied films that were shot there. Gillian Rose's (2007) *Visual Methodologies* provides a thorough overview of varying ways in which visual artefacts can be analysed and interpreted as part of a geographical research project.

*KM*

# 3.3 BODY

When we stop and give it some thought, it seems intuitively obvious that our experience of cities depends upon the manifold capacities of bodies, and the ways of moving, walking, resting, touching, gesturing, sensing, feeling and perceiving they afford us. Yet, when we think about it further, none of these activities is dependent upon our having any sense of what kind of concept a body is or how we might conceptualise this thing. So any effort to discuss the body as a concept must necessarily begin with a kind of qualified admission: bodies are always more than the terms by which they are conceptualised. That said, it remains important to consider how bodies have been conceptualised, and for at least three reasons: first, because how we conceptualise bodies has an important bearing upon what we understand thinking (and, by implication, thinking about cities) to be; second, the matter of conceptualising bodies is a political one, because it involves coming to terms with how bodies register difference; and third, because thinking about bodies involves foregrounding the affective dimensions of urban space.

107

## Conceptualising bodies

To begin it is worth reflecting upon the position of the body within the broader philosophical frameworks and traditions of western knowledge production. Key here is the relation between the body and the activity of thinking. In much of western philosophy the body is defined as supplementary rather than integral to the activity of thinking: René Descartes (1596–1650) is often identified as a pivotal figure in this regard. For Descartes, mind and body are obviously related, but very different insofar as they consist of different substances, one divisible (the body) and the other indivisible (the mind). As a thought experiment, he imagines what it would be like to cut off a piece of his body – a limb, for instance. While this would result in a loss of part of his body, it would not necessarily reduce his capacity to think. If the body can be divided like this without any effect on the mind, it must, therefore, be made of a different kind of substance.

Descartes' affirmation of the difference between mind and body has two important implications. First, it separates thinking (as something which takes place within the mind) from phenomenon like emotion (located within the body). Second, it establishes a split between the rational, thinking subject (the mind) and an unthinking, irrational object (the body). Such divisions form the basis for much of western thinking, including attempts to think about the city. Following their logic means that understanding the dynamics and development of cities should be based upon a kind of rational objectivity that excludes as far as possible the influence of irrational, embodied processes such as emotion. And, in turn, this kind of thinking informs much of urban planning.

While influential, this understanding of the relation between mind and body has been challenged by a 'minor' philosophical tradition in which the body figures in more affirmative and active terms. As a point of comparison with Descartes, the Dutch philosopher Benedict de Spinoza (1632–1677) is particularly significant. Spinoza, like Descartes, affirms the importance of rational thinking. Yet, unlike Descartes, Spinoza does not claim mind and body are two different substances: both are expressions, albeit different kinds, of the same substance. So if Descartes tends to objectify the body, what we get from Spinoza is an understanding of the body as a kind of play of relations and forces: this means that bodies are defined for Spinoza less in terms of what kinds of things they are than in terms of what they can or cannot *do*. A similarly dynamic sense of bodies is found also in the work of a much later thinker, Friedrich Nietzsche (1844–1900). For Nietzsche, western thinking and society needs to reaffirm the body as the source of knowledge and wisdom, even if this knowledge is not produced by the transcendent and disembodied subject which figures in Descartes' philosophy.

A further major challenge to the Cartesian mind–body split comes from feminist scholars. More specifically, feminist thinkers have questioned the tendency to render masculine the rational, abstract, thinking mind, and to feminise the corporeal and emotional. Feminist theorists range from those who argue against any attempt to essentialise the relation between gender and physiology, to those, like Elizabeth Grosz (1994) who argue that bodies are volatile combinations of flesh, fluids, organs, skeletal structure *and* dreams, desires, ideas, social conventions and habits. At the very least, such work complicates any attempt to understand the production of urban space from the position of detached, disembodied objectivity, or to render bodies as passive objects moving within the geometrical grid of abstract space.

# Embodying urban politics

Traditions of thinking about the body have relevance beyond the realm of philosophy. They have also informed attempts to theorise the political, and the kinds of spaces, including the city, in which politics take place. Indeed, understandings of what constitutes a political entity are often framed in corporeal terms: the idea of the body politic exemplifies this. Writing in 1651 the English political philosopher Thomas Hobbes (1588–1679) famously compared the state to a kind of body, 'in which the sovereignty is an artificial soul, as giving life and motion to the whole body' (1985: 81). In this 'artificial body', the officers of the law are the joints, systems of reward and punishment are akin to a nervous system, and wealth and riches its strength. Cities are key sites at which these relations between bodies and politics have been imagined and organized. In ancient Greece, the city came to function as a model for the operation of public life and democratic politics. In many ways, contemporary assumptions about representative politics are modelled on this earlier version of the *polis*. Yet such spaces were not inclusive in the way contemporary democracies are (or at least aspire to be): they operated upon the basis of the exclusion of certain kinds of bodies. Significantly, it was only a rather select number of *men* who participated in this political space. As Giorgio Agamben (1998) has written, the essence of this formulation of the space of politics is more a matter of the kinds of bodies it excludes than the kind of ideal city it purports to construct.

The key point here is that the body is not only bound up in different conceptualisations of the political space of the city: more than this, it is a site at which the political life of urban space is produced, reproduced and contested. Such politics can and have been explored critically in a number of ways. First, the body is understood as a site through which difference is produced and contested. Motivated in part by a critique of accounts that invoke an undifferentiated body as an object of analysis, feminist scholars, including geographers, have revealed how the politics of difference is implicated in the location, regulation, experience and representation of sexed, gendered, raced and classed bodies in urban space. For instance, Linda McDowell (1995) has demonstrated how, in financial institutions based in the City of London, distinctive kinds of embodied performance define the workspace as a masculine and aggressive environment. Within this environment, female bankers only tend to be accepted if they negotiate and in some ways conform to the embodied norms of this workspace.

Second, influenced by the work of Michel Foucault (1926–1984), there is growing recognition of how the body is the object of a range of political techniques and tactics that attempt to regulate and work upon it in ways that are productive of particular forms of behaviour and conduct (see Sennett, 1994; Bridge, 2005). Such claims provide some purchase on the planning and regulation of urban space. Even if urban planning is not always conceived on a grand scale, the idea and ideal of the rationally planned and ordered city problematises bodies as objects with the potential to disturb this order. As such, the emergence of urban order as a political problematic is accompanied by a tendency to police the kinds of activities and bodies that can legitimately take and occupy place in putatively public spaces including the street and the square. In turn, forms of regulation and conduct are complicated by practices that radically alter the capacities of bodies: drinking alcohol is an obvious example, increasing dramatically as it does the volatility of bodies. Regulating such forms of conduct is no less visible and contested today: geographers and other social scientists are engaging with how regimes and systems of surveillance in urban space are becoming ever more elaborate and complex (Koskela, 2000).

Third, even as it becomes the object of techniques and technologies of regulation, the body remains a key site of political transformation, particularly through practices and performances whose focus is the articulation of identity. Parades and marches are obvious and spectacular cases in point. These events can of course be organised around specific political aims or goals, in which case it is often the sheer presence of a mass of bodies in urban space that is of importance. But even events not explicitly organised around a kind of oppositional politics of resistance reveal the ways in which identity is articulated and contested through bodies. For instance, the annual St. Patrick's Day parade in New York can be understood as a particular manifestation and embodied performance of Irishness. While it may cast itself as celebratory and affirmative, recently the limits of this performance of Irishness have been cast in a sharp light by conflicts over the participation in the parade of gay and lesbian groups, something which has been actively resisted (Marston, 2002).

## Embodying urban experience

Foregrounding the body is also important because it provides a way of thinking about and thinking through the spatial aspects of urban experience. A concern with how cities affect experience emerges as a key problematic in writing about urban modernity. For instance, in the late ninetieth century, a particular kind of urban experience came to be symbolised by the *flâneur*: a figure who drifted or wandered at leisure through the crowds of the city, taking particular pleasure from experience of sensuous immersion in these crowds, while simultaneously remaining detached from the throng in a mode of speculative analysis (see Benjamin, 1999). While this figure provided for a distinctive, and distinctively *embodied* mode of urban engagement, it was also one that tended to be overwhelmingly male (although see Wilson, 1991).

This relation between experience and embodiment has been an ongoing concern within studies of urban modernity. For instance, the German sociologist Georg Simmel (1858–1918) argued that one of the key characteristics of modern urbanity was an over-stimulation of sensory and mental life, a development he considered to be negative insofar as it was actually generating new forms of social isolation and emotional detachment (see Simmel, 1997). Equally, the built environments of contemporary cities have also been understood as lacking sufficient stimulus. For instance, the sociologist Richard Sennett has commented upon the apparent contradiction between modern forms of society which seem to 'have so privileged the sensations of the body', and the 'sensory deprivation which seems to curse most modern building; the dullness, the monotony, and the tactile sterility which afflicts the urban environment' (Sennett, 1994: 15–16). For Sennett, one of the key ways in which this sensory deprivation is manifest is through the growing fear of policing, and reduction of tactility and touch.

The French social theorist Henri Lefebvre (1901 – 1991) discusses the relation between the bodily and spatial dimensions of cities particularly forcefully. For Lefebvre, modern, rational urban planning is problematic because it 'overlooks the core and foundation of space, the total body, brain, gestures, and so forth. It forgets that space does not consist in the projection of an intellectual representation, but that it is first of all heard (listened to) and enacted (through physical gestures and movements)' (1991: 200). Any attempt to remake cities must, argues Lefebvre, acknowledge the centrality of embodied experience to the production, reproduction and contestation of urban space.

A corollary of this claim is that thinking through the relation between the body and urban space involves paying attention to details that might ordinarily fall below the threshold of geographic research: ways of walking, taking and gesturing, for instance (Katz, 1999; Laurier and Philo, 2006a, 2006b). Going beyond this, it is also possible

111

for the body of the researcher to become part of process of thinking through the city. Indeed, Lefebvre (2004) argues that the urban thinker should conceive of the body as a kind of metronome through which the rhythms of urban life can be apprehended. Similarly, in a participatory ethnography of boxing, the sociologist Loïc Wacquant (2004) argues for the importance of thinking of and experimenting with the body as vehicle through which to understand the sensual embodiment and modification of experience in cities.

## Affective bodies and urban space

Bodies participate in what might be called the 'infra-structures' of urban experience in ways that do not involve or depend upon contemplative, representational thinking. In this regard geographers and others have recently begun to pay much more attention to the importance of affectivity as a register of urban experience (see Thrift, 2005b). Furthermore there is a growing recognition of how affective processes are differentiated and how this differentiation complicates our understanding of the affective life of cities. The distinction between affect, feeling and emotion is of particular significance here. Following the work of Spinoza, and its subsequent elaboration in the writing of Gilles Deleuze (1989) and Brian Massumi (2002), affect can be understood in terms of a pre-personal intensity of relation between bodies, where bodies do not necessarily need to be human. For example, the play of activity on a football pitch can be understood as a field of affect – a set of relations of continuously transforming duration and intensity. Then, and second, feeling can be understood as the sensed registering of this intensity in a body. The registering of the field of affective relation on a football pitch in bodies – whether players or spectators – can be understood in terms of the process by which affect becomes feeling. Third, and finally, emotion can be understood as sensed intensity articulated and expressed in a socially recognisable form of expression. So the post-event response 'I am overjoyed' by a player to a question posed by a sports reporter can be understood as the expression of emotion.

Cities are obviously sites for the generation of all kinds of affective experience, differentiated by age, gender, class, ethnicity, etc (see Pain, 2001). Affective experience consists in part of distributed atmospheres and intensities of feeling that circulate through and register differentially

112

in bodies and agencies of various kinds. For example, the importance of fear has become only too readily apparent in the wake of the events of September 11th 2001 and the subsequent 'war on terror', whose key sites of intervention have been cities (Graham, 2004a). While generating atmospheres of fear is obviously often the result of very real and all too violent interventions (Thrift, 2007), it can involve also the affective force of as yet unactualised futures, through which the fear of a potential event feeds back with material force into the organisation of the present (see Massumi, 2005). The fact that the materiality of urban space can be reconfigured – through new security measures, spaces of detention etc. – in the face of such potential threats surely makes it all the more necessary that urban geographers continue to develop their understanding of the multiple ways in which the affectivity of cities come to matter.

## Bodies matter more than we think

Lest this become too gloomy, it is worth adding a reminder that the manifold and inventive capacities of the body for all sorts of affirmative investments afford opportunities providing sources of hope, wonder and joy in urban life. These are revealed in a number of ways: in the capacities of bodies to facilitate everyday improvisation and inventiveness through practices such as dance (Malbon, 1999), skateboarding (Borden, 2001), walking (deCerteau, 1984), and even driving (Katz, 1999); in the opportunities bodies afford for cultivating attachments with other forms of sensate life more usually associated with the category of *nature*; and through their facilitation of all kinds of affectively charged association and participation in the public life of urban space exemplified activities like marches, marathons and festivals.

113

The affirmative possibilities of affectivity should also remind us of the excessive quality of bodies: the fact that they will always escape any attempt to capture them in conceptual terms. If to speak of the body is to invoke a complicated set of processes, partly physiological and partly socio-cultural, then it makes little sense to conceptualise the body as a singular, universal entity. Furthermore, bodies are not simply things moving through or within urban space: the ongoing emergence of urban space is a process in which bodies of all kinds participate, metaphorically, materially and affectively. As Spinoza reminds us, such an understanding of bodies should make us less insistent about defining what urban bodies are: rather, we might be better advised to apprehend how the organisation, transformation and experience of cities emerges and depends upon the many things affective bodies can and cannot do.

## KEY POINTS

- Bodies are at the centre of how urban space is experienced, yet much of western knowledge has excluded the body from the business of thinking about urban space. The influence of this can be seen in much of urban planning and governance through the effort to regulate and control bodies and forms of embodied conduct.
- The body is a key site at which the politics of urban space is registered, reproduced, and contested.
- Urban geographers and others have recently foregrounded affectivity as an important element in understanding the relation between bodies and cities.

114

## FURTHER READING

In their paper *Cold Shoulders and Napkins Handed: Gestures of responsibility* Eric Laurier and Chris Philo (2006a) provide an interesting discussion of how the everyday spaces of urban life work (and sometimes don't work) through very ordinary embodied practices and gestures. The work of Linda McDowell provides an important insight into how the identities of bodies in urban (work) spaces are produced through a range of performances and practices. See *Capital Culture* (1997) and more recent work with Adina Batnitzky and Sarah Dyer (2007) on the embodied labours of migrant workers in a London hotel. There is a growing body of literature on affectivity and urbanism, but for an overview of the importance of affectivity in urban life see Nigel Thrift's (2005) paper, *But Malice Aforethought: Cities and the natural history of hatred*.

DMcC

# 3.4 VIRTUALITY

The concept of the virtual is often linked closely with the digital, and the virtual city is often understood as an almost-but-not-quite-real environment facilitated by digital information technologies. Viewed this way, the virtual city is a kind of parallel urban world, a 'city of bits' rather than bricks (Mitchell, 1995), accessible through screens, navigable via keyboards and consoles, and qualitatively different from the real urban environment of houses and streets. The emergence of such urban worlds poses a number of important questions about the structure, organisation and experience of cities. Yet it would be a mistake to restrict any discussion of the concept of the virtual to its association with the digital (Shields, 2003). The virtual is not a creation of late twentieth century computer power: it has been the object of a much longer history of techniques and technologies. Furthermore, the concept has been the subject of philosophical discussion since well before the more recent development of digitally enhanced virtual realities. As such, the concept of the virtual needs to be differentiated into at least three distinct but related meanings: the imaginative virtual, the digital virtual and the temporal virtual.

## Imaginative virtualities

To begin then, what does it mean to claim that the virtual is an element of the imaginative dimension of urban space? A useful point of departure is to consider the experience of reading a literary account of a city (e.g. Joyce's Dublin, Dickens' London, or Zola's Paris) and to consider the nature of this kind of experience. Clearly, the reader's experience of this city is not of an actual place even if some locations and sites in the account might refer to real places: nor does the reader actually occupy the spaces described in the text. Yet if the account is in any way successful, the reader will have a sense, albeit an imaginative one, of being in a particular urban world. This world can be considered virtual because it can be experienced but not in the same way as an actual, physical city. Similarly, memory has something of the virtual about it. The memory of any given urban space is virtual for the simple and obvious reason that no actual physical city is reconstructed in our head. In a manner similar to what happens during the reading of a literary account of a city, memory is generative of an experience akin to being almost but not quite there – the imaginative virtual.

This imaginative virtual is a constitutive element of the experience of urban space, and is something that many non-digital techniques and technologies are designed to work upon and amplify. The history of such techniques and technologies can be traced through a range of practices and sites (see Boyer, 1996b). The most obvious of these is religion, insofar as an almost but not quite real dimension of experience is central to the practice of faith. Furthermore, it is in religion that the idea of the virtual reveals its etymological connection with the idea of virtue. Constructing the conditions for evoking an imaginative sense of the virtuous virtual is part of the architectural design of many of the religious spaces in cities, exemplified in gothic and baroque cathedrals (see Shields, 2003). Of course, churches were also constructed for other, far less noble, reasons: yet insofar as their purpose is in part to allow those present to commune with another world, churches can be understood as architectural machines for producing the experience of virtuality (see Klein, 2004).

116

## Digital virtualities

The key point here is that the revolution in information technology did not so much usher in the space of the virtual: it provided instead for the digitalisation of the virtual as an already existing element of urban experience. This provides a more qualified and nuanced approach to understanding the geographies of the digital virtual, of which at least four aspects have received particular attention. The first concerns the manner in which digital information technologies refigure the spatial relations between and within cities. Here the emphasis is on how the acceleration, increased quantity, and improved quality of information transmission facilitate new ways of organising and connecting urban economies as part of what Manuel Castells (1989) called the 'informational mode of development'. Crucially, Castells' account highlighted the fact that the spatial effects of this new mode of development are registered differentially: not everywhere is networked into the geographies of the digital virtual (Graham and Marvin, 1996).

A second way in which the spatiality of the digital virtual has been explored is through inquiry into the distinctive kinds of spaces of experience it facilitates. Earlier versions of such commentary lamented or celebrated the apparent rendering redundant of real bodies by virtual urban environments. However recent work displays a more sophisticated understanding of how the complex relation between real and digitally enhanced spaces of experience is articulated through sensory registers such as vision (see Boyer, 1996a).

A third way in which geographers are engaging with the relation between the digital virtual and the urban is through the attempt to use the virtual as an object and space of research and teaching. In relation to the former, this can involve the construction of remarkably realistic virtual environments that represent actual spaces, which can then be used to make visible the effects of various future scenarios (such as sea-level change) on cities (see Batty, 2005, 2006; Batty and Smith, 2002,). Using the virtual city as a research or pedagogical environment points to a fourth and final aspect of the relation between the digital virtual and urban space. This concerns the effects of the digital virtual on the socio-political life of the city. Initial critical responses to the emergence and development of technological infrastructures were couched in terms of a spatial politics: they promised to liberate individuals from the constraints of locality by allowing friction-free and democratic mobility on the 'information super-highway'. At the same time, real cities sought to make political communication and representation more democratic by constructing community websites or electronic villages, or by using specially designed software to enhance citizen participation in political decision-making. Yet while laudable, the virtual version of public space often presents a selective and sanitised version of real urban public space (see Graham and Aurigi, 1997; Whyte and Mackintosh, 2003).

This final point is particularly important: it serves as a reminder of the ease with which discussion of the digital-virtual can overstate the novelty and effects of technologies. As Nigel Thrift (1996) has argued, this kind of overstatement rehearses a version of technological determinism in which innovation is radically desocialised, presented as a force acting upon society as if it had newly arrived from another world. The novelty of 'new technologies' is often little more than an incremental reworking of already existing techniques. Nor do digital technologies make the space of interaction within cities any less social than it already was. Thrift demonstrates this point through a discussion of the financial services sector in the City of London, where social interactions (such as face-to-face meeting) and pre-digital representational practices (such as reading) remain central to how the virtual geographies of the spaces of flows are produced and reproduced. As Thrift puts it, the City of London 'looks less like an outbreak of a new and alien information space, inhabited by the plugged-in and zoned-out, and more like a more complex, historically constructed set of "technological" practices' (1996: 1485).

## Temporal virtualities

Many – if not most – of the transactions in the City of London are digitally virtual: they do not necessarily involve the exchange of real objects but consist rather of flows of digital information. But certain kinds of products traded in the City have a quality that points to a third way in which the concept of the virtual can be understood in relation to urban space. These products, such as derivatives, are not simply virtual in the sense of being imagined and/or digital. They are virtual in the sense that the processes or events that determine their value have not yet happened: these processes or events are indeterminate futures. Yet these indeterminate futures become the object of certain kinds of financial decision-making and form the basis of distinctive financial products. So the City of London can be understood as a particularly privileged site for the production of a third kind of register of the virtual: one that can be called the temporal virtual.

The genealogy of this understanding of the virtual can be traced through the work of novelists such as Marcel Proust, and philosophers including Henri Bergson, Gilles Deleuze and Brian Massumi (2002). From the work of these figures emerges a conception of the virtual that can be summarised as follows: that which is real without being actual. What does this statement mean, and more importantly, what is its significance in terms of developing an understanding of cities? The answer to both questions has a great deal to do with how the relation between space and time is conceived. For a thinker like Bergson (1859–1941), writing at the end of the nineteenth and beginning of the twentieth century, the dominant way in which time was conceptualised was through its spatialisation. That is to say that time tended to be understood like another type of space: a kind of container, stretching back into the past and forwards into the future, within which action, events and objects are situated. This conception of time is connected with measurement. For instance, temporal distance from previous generations can be measured by means of years, decades and centuries. Similarly, while exactly what happens in the future remains unknown, we have a sense that whatever unfolds will do so within a temporal container that can be divided up into discrete units. This is the kind of time that underpins the following questions: What age are you? Where do you see yourself in five years? What do you want to be when you grow up? What kind of projections can you make about traffic volume in Dublin 20 years from now?

But this conception of time as a kind of continuous line divided up into units (seconds, milliseconds, nanoseconds, etc.) raises some rather perplexing questions. Assuming that the line can be divided thus, we might well ask how long the present (the 'now' that we experience) lasts? At what point along our imaginary timeline does the present begin and end? This is an important question, because if we cannot identify the points at which the present begins and ends, then what sense does it make to speak of the present as a discrete unit of time or experience? Furthermore, and equally perplexing, what is the status of those things or events that have already passed along the line of time, or which are yet to arrive, positioned as they are somewhere further along the line? We might say that given the fact that they are either in the past or in the future then either they no longer or do not yet exist and therefore can be neither real nor actual. Intuitive as it might be, this claim is rather strange. Given the point just made about the difficulty of defining the unit of the present, we have no way of ascertaining at which point the unreality of the future is transformed into the reality of the present and then on into unreality of the past. We are left with the claim that the present is an infinitely small moment of reality that transforms the unreality of the future into the unreality of the past: the present seems to have a magical capacity to take nothing, transform it into something, before transforming it once again into nothing. Or in other words, in the standard conception of time, the future does not exist, nor does the past, and we have no idea how long the present is, or even if it makes sense to speak of the present as something that lasts any length of time.

Of course, it is possible to counter this claim. What we experience as the past is really a series of representational snapshots – images stored in our heads that provide us with a reservoir of memory into which we dip, sometimes voluntarily and sometimes involuntarily. Similarly, we can imagine the future by constructing images of it within our heads or through various representational media. But this appeal to memory or imagination as merely representations does not really supply a

119

satisfactory answer, for a number of reasons. First, such image-making would have to take place within an infinitely short length of time – in a present whose duration we have no sense of. Furthermore, given the apparent indivisibility and continuity of our experience of the present, how would we know when one unit of the present had finished and another begun? Without this knowledge how would it be possible for us, or our brains, to take another snapshot of the world? And even if such a snapshot could be taken, how would we remember what happened between the snapshots? For surely something must happen between the snapshots. Again, this claim can be countered by the suggestion that memory is more akin to a film than a series of snapshots. But what is film but a series of snapshots projected at a particular speed?

How then can we distinguish between the present as something real, and both the past and future commonly understood to be no longer or not yet real? Bergson's solution to this question is to claim that present, past and future are all real, albeit in different ways. What defines the difference between them is that the present is actual while both past and future are virtual, where the latter is taken to mean *real without being actual*. Moreover, there is no strict division between the actual and the virtual as far as our experience is concerned. The virtual past/future is always being actualised: otherwise our sense of passing through time/space would be impossible. So rather than a kind of window moving along a continuous time-line, through which we take snapshots for future reference of a present that is always passing into the past, the present is an ongoing process of the actualisation of virtuality in which the past and future are always folding into one another.

120

## Virtual urbanities

This understanding of the virtual is becoming increasingly influential in analyses of urban space, particularly those associated with non-representational theories. For instance, near the opening of *Cities: Reimagining the Urban*, Ash Amin and Nigel Thrift announce their intention 'to conceive cities as virtualities, [...] not as being instantiated through replications of the present, but as a set of potentials which contain unpredictable elements' (2002: 4). Put another way, Amin and Thrift are arguing that our sense of the spatio-temporal reality of the city cannot be defined in terms of an endless series of snapshot-like present moments. The reality of urban space is not only a matter of the actual, or of the bare fact of the here and now. As we have seen, this here and now only makes sense if understood in relation to the real virtuality of both the future and the past. For Amin and Thrift then, the reality of urban space consists of multiple pasts and futures differentially actualised in the present in unpredictable and open-ended ways.

While challenging, this understanding of the virtual provides some purchase on the nature and power of certain peculiarly elusive objects of political and economic analysis: those strange futures upon which derivates are based, or those as yet unknown terrorist threats that precipitate the reordering of urban space. Such things do not actually need to exist, but they are real in the sense that their potential effects become part of how the present is organised. Furthermore, and perhaps most importantly, by complicating (and necessarily so) our understanding of what kind of object the city really is, the temporal virtual also complicates how we think about the city, the theories we use in this endeavour, and our expectations about what might happen when we engage in such thinking. For instance, it might mean that the challenge of urban theory is not to get at or build a conceptual model that corresponds to an actual city: the challenge is to develop ways of thinking urban space in terms of a dynamic mix of virtuality and actuality. This does not preclude engagement with the imaginative or digital elements of the virtual: it instead involves thinking about how the real virtuality of the city is actualised through the design, use and experience of a range of technologies including, but limited to, the digital.

121

## KEY POINTS

- While often associated with digital experience, the virtual can be differentiated into three elements: the imaginative virtual; the digital virtual; and the temporal virtual.
- The imaginative virtual is an important element of everyday urban experience, through literary accounts, memories, dreams and desires.
- Recent work within geography has drawn on philosophy in order to foreground the virtual futures constitutive of urban space.

## FURTHER READING

Rob Shields' (2003) *The Virtual* provides an accessible introductory overview of the different meanings of the term. Although well over a decade old, Nigel Thrift's (1996) paper on *New Urban Eras and Old Technological Fears: Reconfiguring the goodwill of electronic things* is worth revisiting as an important qualification of the hyperbolae which tended to characterise 1990s academic writing about the novelty of digital technologies and the kinds of virtual urban worlds they were purportedly opening up. Mike Crang, Stephen Graham and Tracey Crosbie (2006) offer a more recent discussion of the importance of the digital virtual in urban space, focusing on the uneven distribution of IT in everyday urban life. Finally, Ash Amin and Nigel Thrift's (2002) *Cities* provides a provocative vision of what it might mean for urban geography to take seriously the claim that the very material stuff of which cities consist is as much virtual as actual.

*DMcC*

# 3.5 SURVEILLANCE

In *1984*, George Orwell portrayed a relentlessly bleak vision of future society, where an omnipresent authority – 'Big Brother' – watches the community via cameras placed throughout the city. Orwell coined one of the most popular tropes of the twentieth century, a response to the ever-growing tentacles of state surveillance that accompanied the emergence of communist political systems across large parts of the post-war world. Today, surveillance technology is as rife and as stealthy as in Orwell's vision. A range of recording devices put in place by governments, corporations and private citizens spans the city:

> From the road tolling system to the mobile phone call, the camera in the subway station to the bar-coded office door key, the loyalty programme in the store to the Internet usage checks at work, surveillance webs are thick in the city. Yet the aim is not necessarily to catch a glimpse of every actual *event*, though that remains an important goal – so much as to anticipate actions, to plan for eventuality. (Lyon 2001: 54)

However, unlike Orwell's Big Brother, surveillance – especially state-sponsored surveillance – is not omnipresent. It is costly to install and maintain, requires potentially limitless human hours in monitoring, interpreting and analysing its results and – with a relatively fixed perspective on an urban street or landscape – will capture many hours of absolute banality. And yet, surveillance has become a much-researched area in recent social science discourse. This has been fed by real acts of urban terrorism and military warfare, the huge popularity of Michel Foucault's work on governmentality (more simply put as an interest in how states meddle in the lives of individuals), and a growing sensibility of cameras, everywhere, as image-capturing technology becomes democratised in price. There has been a striking *miniaturisation* and *mobilisation* of surveillance technology, including mobile phone cameras, tourist cameras, CCTV and web cams.

## Cameras in the city

Unsurprisingly, much work has focused on the most ubiquitous and obvious type of surveillance technology within the city, CCTV (Closed Circuit Television), a topic of growing interest for scholars across a number of disciplines (see Neyland, 2006: 599–600). Neyland's account of a CCTV system in a British town of 65,000 people reveals the following set-up:

123

> The town contains a CCTV system of 32 cameras connected via a fibre-optic network to a control centre. In the control centre there are 16 small split-screen monitors and 2 large monitors (one to increase the size of the smaller images and one that is mirrored in the local police station). The centre is operated by teams of staff, 24 hours a day. The staff are connected to local police, other emergency services, and local retailers via a radio system. The CCTV system constantly runs time-lapse recorders which compress an image from each camera, every few seconds, onto a single tape. The staff can switch on a real-time recorder if they perceive an event is happening or likely to happen. (2006:604)

For Neyland, this usually mundane activity – which he uses to counterpose against the panopticon or 'big brother' approach – can be theorised as a form of 'spatial accounting', which through a process of verbal discussion between camera operators and police is verified, analysed and 'performed'. Thus the pure, technically captured image is not enough on its own, but is instead formed through professional interactions, 'between

CCTV staff and their managers, between CCTV staff and police officers, and in courts of law' (p. 601). Power relationships pervade this type of surveillance, which underscores a need to focus on how these might be related to gender, ethnicity, class and other forms of social ordering.

However, while this is how the mundane procedure of envisioning occurs, there are strong arguments which place these operations in the context of the interventionist 'police' state. By monitoring landscapes and their users, such as traffic and pedestrians, pavements and building interiors such as banks, airports, casinos and hotels, CCTV has been accused of contributing to wider processes of social sorting within urban space, often justified through an appeal to citizen insecurity over terrorist threats (Lyon, 2003; Coaffee, 2004). While this technology might seem to be oppressive, some suggest that there is a positive side to the outward power surveillance technology has over its subjects:

> It is important to note from the outset that surveillance may be seen here as productive power, and not merely in its more paranoid panoptic guise. Safety, security and social order are all seen by most people as positive accomplishments ... Many surveillance practices and devices are intended to improve city life in significant respects and are welcomed as such. (Lyon, 2001: 53)

Yet the paranoic register has tended to dominate commentary. In his book *Ecology of Fear* (1998), Mike Davis discusses the uses of surveillance technology in the urban environment in regard to the phenomenon of 'smart' or sentient buildings. He argues that such buildings – skyscrapers in particular – are 'packed with deadly firepower' because of the massive amount of hardwiring they contain to operate its sensory system:

> The sensory systems of many of Los Angeles' new office towers already include panopticon vision, smell, sensitivity to temperature and humidity, motion detection, and, in a few cases, hearing. Some architects now predict the day is coming when a building's own artificially intelligent computers will be able to automatically identify its human population, and even respond to their emotional states, especially fear or panic. Without dispatching security personnel, the building will be able to manage crises both minor (like ordering street people out of the building or preventing them from using toilets) and major (like trapping burglars in an elevator). (Davis, 1998: 368)

Such sentient buildings are one part of what Davis labels the 'scanscape', the proliferation of cameras and recording devices within the city's landscape, where even *buildings* are 'sentient', able to anticipate the identity of their inhabitants and even the personal details of the person's requirements (they like the temperature of their room warm, rather than cold, the electric lighting dim rather than bright, their blinds up as opposed to down). Yet surveillance remains tied up in the visual practices of everyday life, where the image on a screen (for example, a security guard's view of a building's lobby) transforms that space into a type of viewing container (Koskela, 2000).

## Surveillance as biometric scanning

However, while surveillance is often used to signify 'sighting' individuals, it has a broader significance in terms of the governance of individual lives. There is a marked variation in the historic ability of states to keep records on the individuals who live in their territory. Having complete records is attractive for a number of reasons: maximising tax returns is one; maintaining internal security is another; and having statistical records to aid in planning infrastructure and future budgets is a third. In Japan, for example, the National Police Authority (*Keisatsuchō*) maintains a network of police boxes in urban neighbourhoods, allowing street-level surveillance, but also conduct a twice-yearly survey of households, their inhabitants, and contents (Wood et al., 2007: 554–5). Such censuses are a key aspect of governance practice, but are conducted with less frequency in other states, partly due to the cost and human resources required to implement them.

What began as fairly primitive attempts to capture the vital statistics of national populations has evolved dramatically alongside the computerisation of records. Many countries have begun to use, or are considering national ID card systems (Lyon, 2004: 1), leading to worries of abuse of an individual's right to privacy. As Graham and Wood (2003) detail, the evolution of algorithmic CCTV (matching facial recognition software and movement recognition to criminal records databases, for example), biometric screening (to allow priority passage at airports for business travellers) and genetic surveillance (which will allow the medical insurance industry to cherry-pick customers) will have significant repercussions for consumer and citizenship rights. Additionally, surveillance gadgetry has become a new commodity, as coveted and convincingly essential to daily life as many of the other luxuries of modern living that are taken for granted.

## Surveillance and the captive consumer

Surveillance in its broadest sense is being enacted daily in the field of **consumption**. Retail theorists have long worked with the concept of geocoding, the mapping of clusters of socio-economic characteristics through postcode (zipcode) based data. Businesses interested in such data have formed a ready market for companies using geographical information systems (GIS), who are able to 'choose targets for coupon promotions, fund-raising appeals, and political pitches [and] also help national chains locate new stores and restaurants' (Monmonier, 2002: 145). Increasingly, the use of store loyalty cards allows retailers to link purchases with individual consumers, allowing both aggregate and individualised targeting of specific niche groups. For example, credit card companies possess sufficient data to make a risk assessment of individuals based on instant access to their borrowing histories, made possible by shared databases. They are then able to offer seemingly endless lines of credit to customers who are able to readily service their debts, providing constant revenue streams (rather than debt repayment per se) for these companies. Surveillance here can be understood as a wide-ranging web of interlocking databases, which without constant government regulation, will increasingly penetrate individual lives, underpinning the contemporary shape of urban retail space, leading to a 'consumer' citizenship (based on one's ability to pay for services), rather than a formal legal citizenship based on equality before the law.

126

## Surveillance as entertainment

It can be seen that surveillance footage has become an increasingly cheap and popular staple of **media** entertainment, particularly in terms of filling the proliferation of commercial television channels. The genre was propelled out of the threat of Hollywood writer's strikes in the United States in the early 2000s, when producers realised that they could make 'a cheap form of niche programming' (Andrejevic, 2004: 2) by voyeuristically filming real people and their reactions to each other instead of using actors on soundstages speaking from well-crafted scripts. Many of these shows are staged in public spaces, or else involve the filming of events that take place in public, meaning that while once confined to television studios, there are now seemingly limitless locations for the setting of such reality shows. For instance, such programmes as *World's Worst Drivers* and *Cops* thrive on the coverage and resolution of legal infractions that take place on public highways, in stores or on the street. Indeed, Biressi and Nunn (2003) argue that 'television and other screen cultures have always exhibited a preoccupation with law breaking and the powerful tensions that it engenders. Both their fictional and journalistic productions offer audiences innumerable variants of crime: its operation, protagonists, and effects' (p. 278). This preoccupation is also expressed in other types of non-fictional television programming, with shows such as *Crimewatch* and *America's Most Wanted*. These shows 'promote themselves as extensions of the law by inviting viewers to help solve crime, provide dramatic reconstructions that allow viewers to "see" the crime take place, offering an experience unavailable to consumers of either news programming or crime fiction' (p. 278). Calvert (2000) describes how these voyeuristic 'reality' programmes resemble

> video vérité … and are voyeuristic because those of us in the television audience never need to interact with the people we observe handcuffed, ambushed, surprised, and exposed on the television screen. The people are simply 'others' out there. We are safely separated and distanced from them. They will not show up, at least not in person, in our living rooms or dens and force us to have discussion with them or share a beer. (Calvert, 2000: 20)

127

# The militarisation of urban space

Surveillance technology has always had a primal link with military operations, from 'spying with maps' (Monmonier, 2002) through to new waves of so-called 'smart' weaponry that are often (erroneously) claimed to be able to pick off individuals or fixed targets from the air. Increasingly, warfare has become urbanised, where the advanced ability to visualise territory – cartography – is fundamental to pin-pointed military bombardments, usually of key infrastructure. On a domestic level, war and terrorism provides economic growth in certain security industries (Marcuse, 2004: 265). Of course, because terrorist activity targets public life, this growth is ultimately most prevalent in public spaces. An example of such security measures are as follows: 'increased policing, K-9 (police dog) bomb teams, sensors to detect chemical, biological and radioactive materials, explosive trace detection devices that scan the air for traces of bomb materials, bomb-resistant trash cans, intrusion alarms, and vehicle barricades' (Marcuse, 2004: 265). These technologies are unevenly spread within the civic spaces of other territories, and few of the residents of Kabul, Belgrade or Baghdad have had such levels of protection.

This requires that the geographer shifts the understanding of territoriality to another dimension, where aerial visual power is the dominating paradigm. Graham (2004c: 21), writing in the context of US air combat in Iraq and Afghanistan, argues the need for a corresponding 'geopolitics of verticality', which can address 'the ways in which the distanciated verticalities of surveillance, targeting and real-time killing confront the corporeal power of resistors to US hyper-power … '. Such a geopolitics 'would need to analyse the ways in which the full might of US military communications, surveillance and targeting systems are now being integrated seamlessly into American civil and network spaces, as well as into transnational ones, as part of the "Homeland Security" drive'.

So, surveillance is a central practice in contemporary urban societies, and while it may aid a sense of citizen security – in the prevention of terrorism, for example, or a reduction in street crime – it also propagates other fears, particularly in terms of loss of individual liberty. As Lyon (2001: 56) argues: 'Surveillance is not simply coercive and controlling. It is often a matter of influence, persuasion and seduction. We are all involved in our surveillance as we leave the tracks and traces that are sensed and surveyed by different surveillance agencies.' Rather than an all-encompassing Big Brother, many urban dwellers may be directly implicated in complex forms of self-surveillance.

## KEY POINTS

- An interest in surveillance studies has grown out of urban terrorism, military warfare and Michel Foucault's work on governmentality.
- 'Spatial accounting' is a process of verbal discussion between camera operators and police.
- Surveillance has a significant role in the governance of individual lives such as crime prevention, internal security and personal finance. It also raises concerns about the invasion of individuals' privacy rights, not least in reality television progrommes.

## FURTHER READING

*Cities, War and Terrorism*, edited by Stephen Graham (2004), collects together 17 chapters exploring the urban impact of terrorism and the urbanisation of warfare in a historical context, and provides a diverse set of contributions to the field. Mark Monmonier's *Spying with Maps* (2002) is an accessible and fascinating overview of the relationship between cartography and surveillance. *Surveillance and Social Sorting,* by David Lyon (2003), an edited collection, considers the social impact of various forms of social profiling carried out in diverse places, such as work-places, homes and transport infrastructure.

*KM/DMcN*

**129**

# 4 Social and Political Organisation

# 4.1 SEGREGATION

At their best, cities are defined by their heterogeneity and diversity. But one of the great paradoxes of urban life is that the apparent freedom of the city is underlaid with complex, often hidden, networks of segregation and exclusion. Some of these are sensible and based on clearly defined and rationally founded principles. Parks are empty of houses, industry and (most) commerce, to create an oasis from the city that surrounds them. Motorways and freeways exclude pedestrians and cycles, who cannot keep to the pace of cars and trucks. Railway and subway lines are separated from the rest of the city, to ensure their smooth running and protect the public. Rainwater is kept apart from sewage in most cities' drainage systems. But many other patterns of segregation are neither natural, nor so obviously rationally grounded. Cities are also segregated by social class, ethnicity and race. They are made up of an intricate mosaic of neighbourhoods and residential areas, which with various degrees of subtlety divide out the wealthy from the comfortably well off, the comfortably well off from the poor, the poor immigrant from the affluent immigrant, the ethnic minority population from the majority population, the bohemian from the affluent social conservative, and so on and on (Knox, 1982; Ley, 1983; Frazier et al., 2003).

132

And there are other more subtle segregations that define the internal morphology of a city. Residential segregation is closely related to occupational segregation. It is not just that different ethnic groups tend to live in different parts of the city, they also tend to do different jobs and work in different places as well. Residential segregation generates differences in where different social groups spend their leisure time – what parks they frequent, what shopping malls or centres they go to, what theatres and cinemas they visit, and so forth. Patterns of residential segregation are also closely associated with the segregation of educational opportunities, with the best schools located in the 'best' – that is to say the wealthiest – areas. Similar patterns are often manifest in other areas of social provision such as health care and childcare. Yet, like so much of urban life, understanding the dynamics of urban segregation is difficult. For one thing, the origins of segregation are multi-causal, segregation is not just an issue of ethnicity, or race, or educational attainment, or of economic position, but of the interaction between these (and other) factors. And, while high levels of class and ethnic segregation can have all sorts of pernicious effects, there are often real advantages in concentrating together as a group. It is often hard to tease out when segregation is in fact chosen from when it is forced upon a particular group (Peach et al., 1981; Peleman, 2002).

# Measuring segregation

Urban geography's initial interest in residential segregation revolved around the question of how to measure and map ethnic residential segregation in urban areas. This work was driven by two concerns. The first was to understand the dynamics behind the sub-division of cities into complex mosaics of land use. The second concern centred on whether certain patterns of segregation might result in a number of unwished-for social outcomes, most noticeably that high levels of spatial segregation among certain ethnic or racial groups would hinder their ability to be a part of mainstream society. Drawing on ideas developed by the Chicago School of Human Ecology in the 1920s and 1930s (see Park, 1925; Burgess, 1925), urban geographers and urban sociologists working in America developed a range of techniques for measuring and mapping the spatial distribution of residential populations within an urban area (Duncan et al., 1961; Tauber and Tauber, 1965; Peach, 1975; Peach et al., 1981).

133

Much of this work revolved around the development of indices that objectively measured levels of segregation. The most widely used of these indices is the Index of Dissimilarity (Duncan and Duncan, 1955; O'Sullivan and Wong, 2007), which:

> ... compares the distribution of two groups, and calculates what proportion of one group would have to move (geographically) to result in an even distribution of both groups across all areas. (Dorling et al., 2007: 46)

So, if, for example, a particular ethnic group had an ID of 40, that would mean that 40 per cent of that group would need to move to create an even distribution of that ethnic group across all areas. The Index of Dissimilarity is calculated using the simple formula:

$$ID = \tfrac{1}{2} \sum_{i=1}^{k} \mid x_i - y_i \mid$$

Where $x_i$ is the percentage of the $x$ population in the $i$th areal unit; $y_i$ is the percentage of the $y$ population in the $i$th areal unit, and the summation is given over all the $k$ units making up the total area, such as a city (Johnston et al., 1986; Shaw and Wheeler, 1985). Other indices include the Index of Isolation (measuring the degree of contact one group has with another), Index of Clustering (measuring the degree to which an area dominated by one group abuts on to another also dominated by that group), Index of Diversity (which measures the degree to which all groups are evenly mixed together) and Index of Centralisation (measuring the degree to which a population is concentrated around a city's core). Indeed, in a review of the literature on segregation Massey and Denton (1988; see also Massey et al., 1996; Simpson, 2007) found no fewer than 20 different indices designed to measure some aspect of residential segregation.

Indices such as these are powerful summarisers of patterns of segregation. But they have several drawbacks. They do not show the actual spatial distribution of segregation within a city – these still have to be mapped. More tellingly, most, and especially the Index of Dissimilarity, are highly scale sensitive. In a classic case, Poole and Boal (1973) studied the level of segregation between Protestant and Catholic communities in Belfast, Northern Ireland. When calculated using Belfast's 15 electoral wards the Dissimilarity Index was 50. This figure rose to a quite disturbing 71 when based on a microanalysis of over 3,000 street frontages. Given these limitations, and aided by the emergence of computer based statistical programmes such as SPSS and Geographic Information Systems, the use of indices like the Index of Dissimilarity have been supplemented by a range of statistical techniques such as cluster analysis, factorial analysis and trend surface modelling, which allow researchers to present a more nuanced picture of patterns of segregation. In particular, recent research has stressed that segregation is multi-dimensional – that it involves not just residential dissimilarity but also patterns of residential isolation, concentration, centralisation and clustering (Massey and Denton, 1993; Massey et al., 1996; O'Sullivan and Wong, 2007; Simpson, 2007).

## Ghettos, slums and ethnic enclaves

Central to much work on segregation, especially in America, is the concept of the ghetto. The term ghetto originally referred to Jewish enclaves in medieval European cities physically separated from the rest of the city. Travelling to America, the term's meaning shifted. Ghetto came to refer to any area dominated by a single ethnic group noticeably different from mainstream society – with the implication that this spatial concentration created a parallel society sharing few of the values and moral norms of mainstream society. Initially the term ghetto was not necessarily homologous with the idea of a slum. Harvey Zorbaugh, in *The Gold Coast and the Slum*, colourfully described slums as:

> ... a bleak area of segregation of the sediment of society; an area of extreme poverty, tenants, ramshackle buildings, of evictions and evaded rents; an area of working mothers and children, of high rates of birth, infant mortality, illegitimacy, and death; an area of Pawnshops and second-hand stores, of gangs, of 'flops' where every bed is a vote. (Zorbaugh in Bell and Newby, 1972: 96)

And while ghettos were often characterised by high levels of poverty, they often also showed degrees of internal social organisation hardly characteristic of slum life. Immigrant ghettos were not simply dumping grounds for society's poorest and most vulnerable. They also provided essential support and aid for new immigrants to gain a foothold in American society. Historical research has shown that most so-called ghettos were remarkably effective in integrating immigrants into mainstream American life. With one exception, none of the ghettos outlined in Burgess's famous 1925 central place model of Chicago existed 40 years later. Indeed, more recent research into the spatial concentration of immigrants in American cities at the end of the nineteenth century and first half of the twentieth century have underlined the porosity of immigrant quarters. Rather than being closed off from the rest of the city these so-called ghettos were in fact characterised by quite significant levels of ethnic diversity. Although there were areas where particular ethnic groups were concentrated, in relatively few cases did one ethnic group make up more than 50 per cent of a neighbourhood. To put things more accurately then, America's early twentieth century immigrant ghettos are better described as enclaves (Philpot, 1978; Lieberson, 1963, 1980; Massey and Denton, 1993).

The exception to the story of gradual ethnic minority integration was the African-American population. And it is the experience of the African-American inner-city ghetto – and the concentrated patterns of poverty and social deprivation that characterise it – that has come to define much of the debate around urban segregation and the so-called ghettoisation of minority populations. Like similarly economically disadvantaged

135

immigrant populations, the African-American population was concentrated in older, inner-city, neighbourhoods. Unlike immigrant groups, however, African-Americans faced extraordinarily deep-seated hostility and often outright violence and discrimination from the existing white population. Intimidation, violent attacks, anti-African-American race riots, lynchings and the destruction of African-American property, in combination with the use of discriminatory legal instruments such as building covenants excluding African-Americans from settling in suburban housing developments, led to the creation of tightly delineated African-American areas that were indeed almost exclusively black (Hirsch, 1983; Massey and Denton, 1993; Sugrue, 1996; Jargowsky, 1997; Fogelson, 2003). Where immigrant enclaves 'served as springboards for broader mobility in society ... blacks were trapped behind in an increasingly impermeable colour line' (Douglas and Denton, 1993: 33).

What is striking – and distinctive – about the African-American ghetto is both the degree of segregation and the level of economic and social deprivation within them. The degree of segregation that defines the African-American ghetto is unique not just because it has much higher levels of concentration than other ethnic minorities in America, but because their segregation is manifested over a range of different dimensions of segregation – African-American ghetto dwellers are highly isolated from other city dwellers, their neighbourhoods cluster together forming one large contiguous enclave, and these neighbourhoods are concentrated in the centre of cities. Under these conditions of 'hyper segregation' and 'hyper-ghettoisation' (Massey and Denton, 1993: 74, 77), it is not entirely fanciful to argue, as the sociologists Douglas Massey and Nancy Denton do, that 'blacks living in the heart of the ghetto are among the most isolated people on earth'. Of course, if this level of segregation were freely chosen then it would not be of much interest. But it is not. As we have seen, it is imposed upon ghetto dwellers. What is more, it is associated with enormous social disadvantage. The African-American ghetto has come to be defined by high levels of unemployment, high levels of welfare claimants and, with that, very high levels of poverty. These high levels of poverty have placed an enormous – indeed unbearable – strain on municipal resources, both increasing the demand for all sorts of public assistance, while undermining the ability of city governments to raise the revenue to pay for public services like schools, libraries, public parks, etc. At the same time, the provision of basic commercial services like banks, supermarkets, chain shopping stores and so forth is highly restricted as non-ghetto businesses see little profit in investing in economically deprived neighbourhoods. As educational and employment opportunities are increasingly circumscribed the possibility of social advancement for those born in the ghetto has become remote. The African-American ghetto has turned into an internally dysfunctional space, 'a human warehouse wherein are discarded those segments of urban society deemed disreputable, derelict, and dangerous' (Wacquant, 2001: 107; 2007; J. Wilson, 1987; Jargowsky, 1997; Frazier et al., 2003; Massey and Denton, 1993; D. Wilson, 2007).

# Suburbia, walls, gated communities: ghettos for the privileged

The long running, and continuing, decline of the African-American inner-city ghetto is often presented as emblematic of the widening social and economic polarisation that many urban commentators see as one of the defining features of contemporary patterns of urbanisation (Mollenkopf and Castells, 1991; Harvey, 2000; D. Wilson, 2007; Wacquant, 1999, 2007). It has also become the focus of a highly politicised debate about the cause and possible solutions to growing levels of urban poverty. While most commentators – as we have seen – stress the importance of economic and political factors in the perpetuation of the African-American ghetto, conservative critics, led by the arguments of Charles Murray (1984) and Lawrence Mead (1986), have argued that the creation of a ghetto 'underclass' is a direct product of a welfare based 'culture of dependency'. This argument, and the welfare reforms inspired by them in the 1990s and 2000s, has led some urban commentators to talk of the emergence of a neo-liberal driven 'urban revanchism' (Smith, 1996: 1; see also Brenner and Theodore, 2002; D. Wilson, 2007) in which aggressive free market policies, combined with brutal policing and other measures, are used to discipline minority populations in an effort to 'reclaim' cities for the 'white middle class'. Nonetheless, perhaps what is most striking about the dynamics of ethnic segregation in wealthy countries outside of America is not how much they mirror the experience of the African-American ghetto (which is, as we have seen, also unique within the American ethnic experience), but the variety of patterns that defines them. While in the UK ethnic minorities are concentrated in the inner-areas of large cities, as they are in many

137

German cities, in France the so-called *banlieue* – areas characterised by high levels of poverty and large immigrant populations – are concentrated on the suburban fringe, as are the similar Swedish *problemområde*. Similarly, the policies for regulating and managing these 'problem' areas are not uniform, just as the degree to which ethnic diversity within so-called immigrant quarters varies significantly (Wacquant, 1999, 2007; Peach, 1996, 1999; Kesteloot et al., 1997).

Despite these ongoing debates on the dynamics of ethnic segregation, within much of contemporary urban geography there is a great deal of unease about focusing on the straightforward mapping and measurement of segregation. In part, this is because it seems to pathologise minority groups for wanting to live in proximity to each other. Another – arguably more significant – source of unease is that focussing on minority groups obscures the extent to which segregation is an issue of the majority culture. It is the majority (in the case of North American and European societies) white population that lives most segregated from other ethnic groups. And furthermore, in the UK, as Daniel Dorling (Wheeler et al., 2005: 1; Dorling, 2007) and his co-workers have demonstrated, middle class and upper-middle class populations inhabit increasingly divergent universes from those below them in the social hierarchy:

> [In the] UK people live in very different worlds. For some, resources and amenities abound; for others life is characterised by deprivation and difficulties even when their need for support is great.

It follows that to grasp the dynamics of contemporary urban development and re-development one of the most pressing needs is to understand the strategies that elite and middle-class groups are employing to manage their contact with the less privileged. Driven by an almost hysterical fear of crime and disorder of any kind, and a desire to protect property values at any cost, innovations such as home owner associations and common interest developments (CIDs, in which residents in a housing development share the cost of maintaining common property), gated communities (where developments are physically gated off from the rest of the city), downtown business improvement districts (BIDs), suburban 'edge city' office developments (located well away from older, 'crime ridden', inner-city areas) and enclosed suburban shopping malls, are reconfiguring urban space in profound and often disturbing ways.

These innovations, many of which sound innocent enough in themselves (see Webster, 2006), are creating an almost uniquely

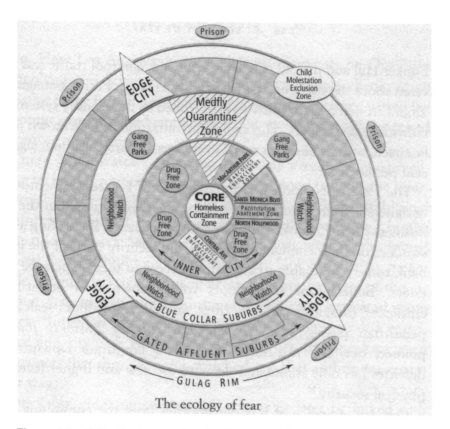

The ecology of fear

**Figure 4.1** 'Mike Davis reapplies the Burgess Chicago model to his Los Angeles-based 'Ecology of Fear'

divided urban world: a city defined not by its meeting places, but by its walls and its enclaves (Marcuse, 1997; Davis, 1998; Soja, 2000; Low, 2004). Of course, many of these developments had been prefigured in earlier periods of urban development. The extraordinary expansion of suburbia in post-World War Two America, for example, was fuelled in no small part by racist anxieties about inner-city crime and disorder, and was structured in all sorts of ways to exclude unwanted minorities (Jackson, 1985; Fishman, 1987; Self, 2003). And contemporary gated communities with their walls, private roads, beautifully landscaped

139

gardens and – in the more exclusive cases – golf courses, swimming pools and gyms, echo earlier residential developments in eighteenth and nineteenth century London and elsewhere.

Nonetheless, four elements are particularly striking about the rise of contemporary 'privatopias' (McKenzie, 1994):

1   *Their growing pervasiveness.* They are an increasing prominent portion of new residential developments.
2   *How little they connect to the immediate city around them.* The use of gates and security guards means there is little porosity between the new enclaves and their neighbouring areas.
3   *The degree to which they are part of an emerging urban geometry where the well-off have no need to interact with the less affluent.* The combination of residential enclaves, automobile oriented edge city office developments and suburban shopping malls, in combination with the private automobile, has created a hermetically sealed network of environments beyond which the privileged need never venture.
4   *The degree to which the patterns of segregation they generate show a remarkable convergence between cities of the so-called Global North and the Global South.* The degrees of segregation that have long been associated with cities like São Paulo and Cape Town, are now increasingly seen in cities like Los Angeles, Houston, New York or London (see Caldiera, 2000; Lemanski, 2006a, 2006b).

## KEY POINTS

- Segregation has been of interest to urban geographers since from the very foundation of the sub-discipline.
- Segregation is multi-casual and multi-dimensional.
- High levels of residential segregation can be corrosive, but ethnic enclaves can also offer essential resources and assistance that facilitate social mobility.
- The contemporary African-American ghetto represents an almost unique case of extreme, multi-dimensional, segregation.
- Middle class and elite urban populations are increasingly seeking to restructure urban space so they do not have to come into contact with people of lower social status than themselves.

Douglas Massey's and Nancy Denton's (1993) *American Apartheid: Segregation and the Making of the Underclass* is a comprehensive account of the causes and the consequences of the hyper-segregation, and hyper-ghettoization, that define contemporary inner-city African-American neighbourhoods. Nan Ellin's (1997) *The Architecture of Fear* provides a series of short, accessible, essays exploring how western cities are being rebuilt by elite groups in response to a heightened sense of fear and insecurity. Teresa Caldeira's (2000) *City of Walls: Crime, Segregation, and Citizenship in São Paulo* is an in-depth study of the relationship between fear of crime, race and the emergence of fortified urban enclaves in São Paulo.

*AL*

# 4.2 URBAN POLITICS <sup>141</sup>

The field of urban politics has never been easy to define. For one thing, it is not limited to human geography. Indeed, it is probably true to say that the disciplines of urban sociology, political science, urban planning, real estate finance and economic development have had more to say about the structure of political activity in cities than have geographers. However, by using spatiality, scale, and place identity as key analytical concepts, geographers have provided a distinctive set of perspectives on how power is effectively urbanised.

The arrival of globalisation both as theory and practice has challenged the assumption that urban politics is determined by scalar hierarchies of size-based territorial states: the city council beneath the regional assembly, beneath the central government, beneath the European union, for example. This has a subtle implication for how the 'local' is framed. For Amin, 'Instead of seeing local political activity as unique, places might be seen as the sites which juxtapose the varied politics – local, national and global – that we find today. What matters is this juxtaposition' (2002: 397). In other words, cities are sites of networked practices which are distanciated, stretched over miles and miles, and may be conducted by people who have never met face-to-face, and with completely dissonant time horizons as to how long they wish to interact with a specific place or locality. For theorists such as Kris Olds (1995, 2001), urban redevelopment politics in cities such as Vancouver is dictated partly by the presence of transnationally networked actors such as property developers, who can mobilise their financial power and knowledge of local markets to outmanoeuvre community opposition. Local actors can thus combine with outsiders to enhance their capacity to govern, whether through greater financial liquidity, or greater knowledge to exploit particular local situations, for example.

With this tension between the local and the non-local, inside and outside, in mind, the following discussion briefly outlines three key issues in urban politics: the role of structure and agency in urban politics; the construction and contestation of place identity; and issues of representation and performance.

## 142 Cities as political actors: structure and agency

One key question is that of who speaks for cities. Critical geographers and sociologists have often argued that business lobby groups aim to paint a picture of a cohesive, unified city, masking the fact that hierarchical, exploitative working conditions may predominate in a locality. The locally dependent hierarchies which dominate the political lives of cities have been theorised as growth coalitions (Logan and Molotch, 1987), which emphasised the importance of political networks, both formal and informal, between actors with a shared interest in maximising economic growth in a locality. Cox (1993, 1998) has argued that these groups have significant commitment to particular places, either through long-established customer bases, major property holdings, trained workforces and long-developed relationships with local suppliers. Alternatively, they may be major civic institutions such as universities, sports teams or newspapers, which – while having a public function – nonetheless operate on profit-making principles. These institutions have often clashed with 'quality of life' groups, particularly environmentalists and homeowners, who seek to preserve amenity (and, often, the values of their homes). Business elites seek growth for the generalised increase in profitability gained from resale, development and rent, and municipal councils seek tax revenues which are derived from these profits. More recently, such approaches have been criticised for being excessively localistic (ignoring the structured power of central government or major corporations, for example) or voluntaristic (e.g., Jonas and Wilson, 1999; and specifically Jessop et al., 1999), endowing local actors with excessive power or capacity to act (see Stone (1989) on regime theory). Such a backlash, which has seen theorists focusing more on the structural constraints placed on actors due to macroeconomic restraints, is often taken as an indication of the shift in the governing principles of national economies.

One of the biggest challenges to this world view came from the sweeping inroads made by neo-liberalism in the world economy (e.g., Kipfer and Keil, 2002; Swyngedouw et al., 2002; Weber, 2002). Critical geographers have been enthusiastic about the explanatory power of this paradigm. For example, Brenner and Theodore's discussion of 'cities and the geography of "actually existing neoliberalism"' is very explicit in its attempt to open up what they call the 'spaces' of neo-liberalism. Here:

> Cities – including their suburban peripheries – have become increasingly important geographical targets and institutional laboratories for a variety of neoliberal policy experiments, from place-marketing, enterprise and empowerment zones, local tax abatements, urban development corporations, public–private partnerships, and new forms of local boosterism to workfare policies, property redevelopment schemes, business-incubator projects, new strategies of social control, policing and surveillance, and a host of other institutional modifications within the local and regional state apparatus … the overarching goal of such neoliberal urban policy experiments is to mobilize city space as an arena both for market-oriented economic growth and for elite consumption practices. (2002: 368)

143

While this rather broad-sweep take on the city misses many of the subtleties of how these changes are mediated in specific contexts, there is no doubting the ideological and regulatory changes that have occurred in many cities in the last two decades. The impact of these changes can be seen across the board, from the precarious working conditions of office cleaners (Herod and Aguiar, 2006), to the competitive logic of waterfront redevelopment (Swyngedouw et al., 2002), to the commodification of arts policy seen in the encouragement of cultural and creative industries (Peck, 2005) to the impact of multinational leisure and brewing corporations on the night-time economy (Hollands and Chatterton, 2003).

# Place, identity and contestation

Giving cities the status of agents has been quite common, magnified in debates around urban entrepreneurialism (Hall and Hubbard, 1998; Harvey, 1998a). Here, territorial governments are seen as firms, engaged in an inter-urban competition with other city governments. These councils actively market strategic sites, quality of life statistics, cultural facilities and labour skills to inward investors. Finance, taxation and capital investment are key issues in a deeper understanding of urban power: 'Despite their snooze value, city budgets are important political documents because they represent a constant struggle over who should pay for and who should receive services and programs' (Judd and Swanstrom, 2006: 330). Understanding such budgets requires some skill, not always possessed by non-specialist urbanists, and a fair degree of persistence. It is also difficult to generalise across countries, as some states such as the US have well-developed forms of local financing, often revolving around the issue of municipal bonds, property taxes, and government aid (Weber, 2002). This has led to the rise of fiscal 'enclaves', patches of urban territory that have separate financial rules, such as districts redeveloped under tax increment financing (TIFs), business improvement districts and development corporations (Weber, 2002; Swyngeduow et al., 2002).

144

There has been a strong emphasis on charting the impact such policies have on place identity, and particularly the impact of manufacturing decline on community politics (e.g., Merrifield, 1993a, 1993b). This has reshaped conventional party politics, where the governing parties of cities such as Barcelona (McNeill, 2003a) or Manchester (Quilley, 1999, 2000) have consciously reshaped their development strategies to attract 'knowledge workers', urban events such as major art exhibitions, sports events and business tourism. This has not been confined to the west. The growth of high technology corridors and nodes in Malaysia has had a significant impact on indigenous groups (e.g., Bunnell, 2004a).

However, it would be a mistake to associate the practice of politics only with state actors. Good examples include Anderson and Jacobs' (1999) study of residential activism in 1970s Sydney; the racialisation of urban politics in the construction of, for example, Chinatowns (Anderson, 1988, 1991); the role of trade unions in community politics (Wills, 2001); or the renegotiation of citizenship associated with AIDS activism (Brown, 1995, 1999).

Nonetheless, the problem with a relational urbanism is that the static, sedentary analysis of territory can be neglected. Theorists have sought ways in which the continuities and genealogies of place can be explored, which allow the specifics of place-based politics to be discussed, while still sensitive to the macro-systems that affect urban politics. There have been some fine examples of this: Allen et al.'s (1998) multi-layered account of the transformation of the highly-urbanised Southeast of England during the 1980s and 1990s; Jacobs' (1994) account of the 'anxieties' of colonial urbanism in contemporary globalisation narratives; Cronon's (1991) study of the integration of city and countryside as a cohesive system in nineteenth century Chicago; Gandy's (2002) study of urbanisation and nature in New York; and Harvey's (2003) grand narrative of spatialised class struggle in nineteenth century Paris. Of these authors, only Cronon is a professional historian, and it is interesting to speculate how the disciplinary perspectives offered by geographers would provide distinctive accounts of urban change in particular localities to those written by historians. A key issue here is bringing out the 'contingencies' of urban politics, the idea that – contrary to what the neo-liberal theorists may argue – policies and practices vary widely from place to place.

Contingency is probably most easily seen in societies in transition from one social and political order to another, as Karen Till has suggested in the context of Berlin:

> Central to the ways that people create meaning about themselves and their pasts is how they expect places to work emotionally, socially, culturally and politically. How do people make places to delimit and represent time (past, present and future)? How, in turn, do these places define social relations? Often a dominant set of culturally place-based practices – what Linda McDowell [1999] calls 'regimes of place' – comes to define how people think about a place's location, social function, landscape form and aesthetics, about international commemorative display, and even personal experiential qualities. But those regimes of place may be questioned during times of social and political transition. (Till, 2005: 11–12)

Globalisation has had a far-reaching impact on Chinese cities (e.g., see the collection of papers in Wu, 2006). In South Africa, the restructuring of the post-apartheid state has given urban politics there a particular specificity (see Robinson's (1998) study of the renegotiation of political identity in the transition to democracy). In Poland, along with other post-socialist economies, workers' movements have found it difficult to sustain wages and rights in the face of a growing neo-liberalisation of the economy (Stenning, 2003).

# Mayors and the imagined geographies of cities

Urban governance – a self-conscious differentiation from 'local government' – has been an important mechanism by which powerful by agents working through multi-scaled states are able to manipulate the political process. This debate has often been framed in terms of structural, macro-economic forces (interest rates, central government planning policy, national and worldwide consumption trends), and has often constructed the governance of urban economies as being strictly curtailed by competition between places (Peterson, 1981). And yet, it would be a mistake to ignore the significance of individual, personalised forms of political and civic leadership. This can even be seen as performative, an idea contained in the more popular concept of 'statecraft'.

The figurehead of city leadership is, of course, the mayor. As 'first citizen', mayors are often associated with political parties, yet many of the most successful mayors are often those who are able to speak 'for' their city. Rudy Giuliani, for example, while pursuing a neo-liberal political agenda, was often seen as being outside of the mainstream of the national Republican party. Furthermore, mayors are often crucial in articulating the interests of their cities to external agents, be they national governments or major public and private investors. This relationship is cemented by a performative relationship to mayoral governance, encapsulated in three inter-related roles. First, mayors 'embody' cities. Many are born in the city they represent (and here representation has a double meaning), and will have some sort of relationship to the essentialised characteristics of its inhabitants (often accent, sense of humour, or 'inheritance' of attributes of an idealised predecessor). Here, they communicate with their voters/fellow citizens. Second, they will act as animator of city space. Rather than pursuing an abstract notion of territory, mayors often strive for visibility in the everyday life of the city, especially at times of crisis (consider Rudy Giuliani's presence around the site of the World Trade Center on and after 9/11). From presiding over ceremonies to high profile public walkabouts, they will seek to become 'everyman', in experiencing the city's urban life (and we should consider the gendered nature of such terms, including the more general idea of the 'city fathers'). Third, they are likely to provide some sort of narrative in their press conferences and public appearances about the immediate past, present and future of the territory they represent, shaping and responding to a public discourse concerning crime, fear of terrorism, the economic climate, etc. (McNeill, 2001a, 2001b).

To conclude, recent years have brought a reconsideration of how the terrain of urban politics is conceptualised. On the one hand, the impact of globalisation theory has prompted many scholars to think through cities as being constituted by flows of resources, exchange rates, migratory patterns and investment decisions taken *elsewhere*. This complicates territorially based understandings of cities, and has been one reason why the term 'local government' (with its assumptions of fixed boundaries, and some degree of political autonomy) is being replaced by urban governance (with an assumption that decisions are strongly influenced by non-state actors such as major firms, and that political practices take place outside the administrative boundaries of the city). We can also trace the growing significance attributed to political agency as being performed, either in public view or through private association, as having material form through discourse, and as avoiding the understanding of the 'state' as being a monolithic, unchanging, 'black boxed' institution. The insights of cultural geography on these matters have enriched the study of urban politics.

147

## KEY POINTS

- Cities cannot be understood as cohesive entities, with singular, common interests. Rather, their governance and internal wealth distribution is highly politicised.
- Urban theorists argue that there has been a transition in governance forms over the last 30 years, with a shift towards neo-liberal forms of urban governance, and a growing impact of transnationally networked elites such as property development firms on city politics.
- Cities are represented and performed, and actors such as mayors have considerable power to communicate specific visions or discourses of what a city should be, has been and will be in the future.

## FURTHER READING

One of the best-known books on urban politics is *City of Quartz: Excavating the Future in Los Angeles,* by Mike Davis (1990), a controversial foreshadowing of the riots that would engulf that city in the early

1990s. Richard Walker (1996) provides a biting critique of the restructuring of San Francisco and the Bay Area over successive decades In *Another Round of Globalization in San Francisco*. Jonas and Wilson's (1999), *The Urban Growth Machine: Critical perspectives, two decades later* offers a wide-ranging discussion of urban politics and theory, though tends to focus on the US and UK. For a deeper cultural geography of the governance of a key space in the contemporary economy, Luisa Bialasiewicz's (2006) paper, *Geographies of production and the contexts of Politics: Dis-location and new ecologies of fear in the Veneto città diffusa*, provides a fascinating counterpoint to the more structural readings of urban politics, conveyed with a deep sensitivity to local context. Anderson's (1988, 1991) accounts of the racialisation of Chinese ethnic groups in nineteenth century Vancouver provide an important basis to the processes of ethnic segregation that have been repeated in cities worldwide.

*DMcN*

148

# 4.3 COMMUNITY

Community is a word with a pleasant, warm, tone. It suggests friendship, connection, cohesion, mutual support. Cups of tea over the back fence, friendly gossip, neighbours looking out for each other. Warm pints of bitter sipped around a pub fireplace, soccer mums cheering for their kids, PTA meetings, church fêtes. Community conjures up images of a world in which things fit together. It is something that just about everyone across the political spectrum is in favour of. And who could be against it? Who would not want cohesion, social warmth, people caring for each other? Yet for all its apparent simplicity community is one of the most ill-defined and argued about concepts within the social sciences. As early as the 1960s Margaret Stacey (1969: 134) was arguing that 'as a concept "community" is not useful for serious sociological analysis'. And this is a sentiment that many urban scholars share (see for example Sudjic, 1991; Harvey, 2000; Thrift, 2005b). Nonetheless, its use continues. And arguably, with the rise on work on social capital and contemporary debates around sustainable cities, it has become more, not less, central to urban debates over the past 40 years.

# What is community?

So, what, conceptually speaking, is community? If there are almost as many definitions of community as there are people who write about it, it is nonetheless possible to outline three principle ways in which the term is used:

1 *Community as place or neighbourhood.* In its most common usage community simply refers to the population – and the interconnections of that population – in a particular area. One speaks about the Islington community, the Kreuzberg community, the Portswood community, the community of Palmerston North, and so on. Often implicit within this definition is some sense that through the fact of their shared environment the people of a place also somehow hold a set of shared values about the proper ways to behave, the proper way to lead a life, and so forth. This, of course, disguises the fact that each of these communities themselves is made up of a tapestry of sub-communities of ethnicity, race, class and religion (to name just three key axes of differentiation). These sub-communities may, in fact, have little in common with each other apart from the fact that all exist side-by-side in a particular area.

2 *Community as a set of shared values, practices, and ways-of-being-in-the-world.* This brings the discussion to the second usage of community, community as a set of common characteristics, social practices, values and beliefs, among a group of people. Prominent examples of the kinds of groups who fall under this usage of community include religious groups, ethnic groups, and class based communities. Such communities may be defined by a particular place, but that is by no means a prerequisite for their existence. Similarly, although this use of community might usually be thought of as being attached to clearly defined and identifiable groups – the 'Muslim community', for example, or the 'gay community', or the 'elderly community' – its boundaries can also bleed out into forms of identity that are chosen and which may constitute what appear to be only a relatively trivial part of a person's life. A central characteristic of this notion of community is that the values and beliefs

149

shared by a group, and which, thus, are in a certain sense constitutive of the community, are generally 'taken for granted', they exist as a kind of shared sentiment, or what the French sociologist Pierre Bourdieu (1977) called a *habitus*. That is to say, while there may be debate within such communities about the finer points of a particular issue of belief or behaviour, and there may be formal systems of rules that define group behaviour and membership, these are underpinned by an implicit and emotionally deep rooted shared understanding of how the world works.

3 *Community as shared interests*. A final, common definition of community is of a group of people that are defined by a shared set of interests. Here the threshold of commonality is lower than for the preceding definition of community. In many cases it may well be that only a relatively narrow aspect of a person's life is involved. Indeed, most people are members of a diverse range of communities of interest. They might be a member of an occupational association, *and*, say, a residents association, *and* a member of a social pressure group, and so on. What is important to stress about this third usage of community is that unlike community as a set of shared values, practices and ways-of-being-in-the-world, the defining characteristic of this form of community is that it is primarily based upon conscious, reasoned, joint-action with other people. A pure community of interest is not based on emotional, or mutual affection, but on a clear-eyed assessment of the benefits to be gained by associating together as a group. Or put more simply, people come together to form part of a community of interest because it is in their best interest to do so.

We could say a lot more about community. But rather than delving ever more deeply into the concept itself it is perhaps more useful to shift the focus a little and examine how the concept of community has been put to use in urban geography and urban studies.

## Community lost

Having started by trying to define community, it might surprise the reader to discover that the most significant feature of much of urban studies' – and with it urban geography's – early concern with community focused not on community at all, but on its absence. Nineteenth and early twentieth-century social thinkers drew a sharp distinction between the emergent 'modern' societies that confronted them, and the supposedly traditional feudal societies from which that modern society had been born. The French social philosopher Émile Durkheim (1947), one of the founding pillars of modern social theory, argued that the movement from feudal to industrial societies was characterised by a transition from a social world built on a 'mechanical solidarity' of custom, obligation and emotion, to an 'organic solidarity' rooted on the interdependencies of a highly variegated, individualised, division of labour where 'community takes smaller toll of us'. And Ferdinand Tönnies (1963), in *Community and Society,* perhaps the most influential book on the transition from feudalism to an urbanised, industrial, society, was convinced that a narrow instrumentality defined social bonds in modern society.

Both Durkheim and Tönnies' accounts were organised around mapping the decline of 'traditional' community on to a rural–urban, pre-modern–modern, dichotomy – rural society stood for tradition and community, urban society for modernity and individualised society. In drawing this division Durkheim and Tönnies provided intellectual force to a deep-seated cultural prejudice among European and North American intellectuals and social commentators who saw in the rise of mass urbanisation the erosion of all that was good and noble (Glass, 1972; Williams, 1973). (The philosopher Jean-Jacques Rousseau called cities 'the abyss of the human species'). This was a theme that was taken up and elaborated by early twentieth-century urban reformers in the UK, North America, and elsewhere. Critics like Patrick Geddes, Ebenezer Howard, Raymond Unwin, Fredrick Osborn, Lewis Mumford and the Regional Planning Association of America argued that the industrial city was a cancerous social form that precluded the development of real community (Miller, 1989; Hall, 1996; Welter, 2002). In the words of Mumford, the industrial city was nothing less than 'the crystallisation of chaos'. A world defined by 'a raw, dissolute environment, and a narrow, constricted, and baffled social life' (Mumford, 1938: 7, 8).

# Community saved, community liberated

The problem with these accounts of modern cities as graveyards of community – popular and pervasive though they are – is that they are quite simply wrong. Modern, industrialised, societies, to say nothing of modern (or indeed post-modern) cities, are different to traditional societies (and cities) in all sorts of ways. But this does not mean that they are bereft of community. Far from it. As the American urban critic Jane Jacobs (1961) wrote in a seminal critique of Howard and Mumford, the problem of cities is not whether they have community or not, but how the community that is there is organised. Once urban researchers began to start looking for community rather than theorising about its absence, they found it in abundance. Neighbourhood studies like William F. Whyte's (1943) *Street Corner Society*, Michael Young and Peter Willmott's (1957) *Family and Kinship in East London*, Herbert Gans' (1962), *The Urban Villagers*, and (1967) *The Levittowners* and Elliot Liebow's (1967) *Tally's Corner* – to name just a few prominent examples – demonstrated the ongoing communal vitality of cities.

One way of interpreting these studies is to suggest that the intimate relationships that defined traditional societies were in fact carried into the modern, industrial, city. But this simple dualistic split between traditional community and the impersonal ties of modern society is not one that holds up to scrutiny. It was not just that community had been 'saved' from the corroding effects of modernity. Community has also been transformed and remade in all sorts of interesting ways. Indeed, if we start thinking about cities as vast theatres of connection, it becomes possible to see that modern urban life has in fact generated a great variety of novel forms of communal association (Fischer, 1982; Wellman, 1998; Amin and Thrift, 2002).

Some of these forms of association, as with the neighbourhood-based studies of Whyte, Young and Willmott, and Gans, remain tightly place based. Many others describe communities that have been 'liberated' (Wellman, 1979: 1206) from the parochialism of place. Barry Wellman (1979; Wellman and Leighton, 1979; Wellman et al., 1988) and Claude Fischer (1982) describe how community for many urbanites is defined through personal networks of friends, co-workers, neighbours and kin – networks that often do not have much spatial overlap. More recently, scholars interested in the lives of international migrants have shown how many migrants have constructed 'transnational communities' (Kearney, 1995) rooted in the place of arrival *and* the migrants' homeland. Indeed, the anthropologist Michael Peter Smith believes the dynamics of these trans-national social networks are so pervasive we can speak of the emergence of a kind of **transnational urbanism** (Smith, 2001).

# Imagined communities, community and exclusion, light sociality

Having discovered that community has not been lost, that it is not necessarily bound to place (and in fact may be entirely unbound from place), what is left of community for urban geography to explore?

## Imagined Community, Mass Communication, and Mass Consumption

Well first, urban geography can explore how community is imagined. Perhaps the biggest conceptual challenge to recognising that cities could in fact maintain and nurture community was the sheer fact of their size. Cities represent an enormous problem of social coordination. How can such enormous numbers of people formulate any kind of meaningful community? If community is thought of as being principally about direct fact-to-face interaction then the idea of cities as repositories of community seems an absurdity. But if we recognise that all communities beyond the smallest involve an imaginary dimension that allows its members to see beyond the bounds of their immediate experience, then it becomes more straightforward – indeed logical – to think about community in such large scale situations. In many societies religion, with its complex rituals functions to integrate disparate social groups into a larger whole; in contemporary cities these pre-existing modes of integration have been overlaid with a range of technologies of mass communication – newspapers, magazines, poster hoardings, cinema, radio, television, to name the most prominent examples – that in all sorts of ways facilitate the imagining of community and connection (McLuhan, 1962; Anderson, 1983; Morley and Robins, 1996).

153

Perhaps the earliest, and certainly initially most powerful of these technologies of mass communication, was the mass press – daily and weekly newspapers and magazines. As the cultural historian Benedict Anderson stressed in his seminal book *Imagined Communities* (1983), mass literacy, combined with cheap mechanical printing, transformed the possibility for constructing connection between large numbers of people. While Anderson's work focused on the relationship between the emergence of the newspaper and nationalism, a parallel process shaped urban life. As early as the 1920s the influential Chicago School writer and former city journalist Robert Park had argued that:

> The growth of great cities has enormously increased the size of the [news-paper] reading public. Reading which was a luxury in the country has become a necessity in the city. In the urban environment literacy is almost as much a necessity as speech itself. (1923: 274)

Park stresses that all newspapers are essentially 'devices for organiz-ing gossip' (p. 277), but newspapers and other forms of mass communi-cation do much more than just facilitate the spread of information and 'gossip'. The simple practice of newspaper reporting lends a certain weight to everyday urban experience. Precisely because newspapers are produced so regularly much of what they report focuses on the tex-ture of everyday existence. Along with sensational ruptures in everyday events – fires, accidents, accounts of criminality, exceptional luck, or ill fortune – they narrate the 'natural' rhythms of a city – its festivals, sporting and political events, and so on (Anderson, 1983: 7; Lewis, 1993; Fritzsche, 1996; Lindner, 1996; Schwartz, 1999).

## Community, Exclusion, and the Good City

Of course to argue that mass communication helps form community is to recognise that community is not spontaneous (or not only sponta-neous) but is also 'constructed' (Robins and Morley, 1996; Latour, 2005;). The stories that newspapers tell, or TV news crews film, or the movies that are made, are put together through a complex amalgam of editorial decisions, technological constraints, and production conven-tions. Indeed, a great deal of recent urban writing has focused on how this mediatisation of urban life has turned cities into a kind of mass spectacle (cf. Debord, 1994; Merrifield, 2002). Michael Dear (2000: 8) describes how the form of postmodern cities like Los Angeles is being 'increasingly determined by the demands of spectacle and **consump-tion**'. While in a similar vein David Harvey (1989a) has written of how the manufacturing of urban spectacle has become a key strategy for capital accumulation (see **architecture**).

Although a prevalent trope in much contemporary urban writing (see Brenner and Short, 1996; Hall and Hubbard, 1998; Dear, 2000; Lees, 2002; Smith, 2002; Theodore, 2002; Hall, 2006; Silk and Andrews, 2006) this argument is problematic. It draws heavily upon the 'community lost' narrative. At the same time it adds a narrative of false consciousness, suggesting that where people do feel a sense of bonding and common connection in the postmodern consumer-oriented contemporary city, it is manufactured and superficial. Or put another way, it is not real community. Nonetheless, this strand of writing does point towards a further way urban geographers are engaging with community. This is to examine the ways in which the term community itself is used, and employed, by members of the public, politicians, the media and urban policy makers (see **urban politics**).

Thus, writers like Mike Raco (2005) and Rob Imrie (Imrie and Raco, 2003) have explored the ways community is used as a rhetorical device in the making of urban policy. The trope of community is employed as a way of both mobilising people in an area in aid of certain policy objectives, and as a way of demarcating what is not considered acceptable. As Imrie and Raco write:

> ... one consequence of 'government through community' in British cities is likely to be the creation of new social division ... Indeed, the potential for people and/or places to be labelled 'inactive' (or dependent or deviant) is part of the policy design which is being applied by the Labour government to the 'renaissance' of Britain's cities. (2003: 6)

Of course, it is not just politicians and policy makers who use community in this way. Drawing on psychoanalytic theory, Steven Pile (1996) has explored the ways that the popular media articulate shifting notions of normalness, belonging and supposed deviance, in the ways that it invokes a sense of 'we-ness' in its reporting on issues of crime, race, ethnicity and sexual difference. And in *Geographies of Exclusion* David Sibley (1995) demonstrates how community can often be extremely defensive. In many cases assertions of local community cohesion are organised around the expulsion of those that the majority defines as 'dirty', 'un-hygienic' and 'impure'. Indeed, in many American cities, community has come to be defined through the presence of walls, gates and security guards designed to keep the confusions of the city outside from intruding on the tranquil atmosphere inside. An unhappy state of affairs explored in Setha Low's (2004) *Behind the Gates: Life Security, and the Pursuit of Happiness in Fortress America* (see **segregation**).

## Passion, Light Sociality and Everyday Connections

These arguments about community and social exclusion tell us a great deal about the dangers and limitations of community. They act as a warning to the great majority of politicians and policy makers who see community – and the associated notion of social capital – as a panacea for solving many of the problems facing the contemporary city. Nonetheless, they do not tell us very much about the ways that connection does actually occur within cities. As we have seen, cities teem with connections, with all sorts of sociality and interaction, some intimate, some anonymous, some formulaic, some enduring, some fleeting. A third way that urban geographers have explored community is – somewhat paradoxically – to try and think beyond it. Rather than considering the ways that certain interactions add up to something larger such as community, writers like Ash Amin (2007), Nigel Thrift (2005; Amin and Thrift, 2002), Derek McCormack (2003; Latham and McCormack, 2004, 2008), Alan Latham (2003b, 2004, 2006), and Eric Laurier and Chris Philo (2006a, 2006b) have focused on the small interactions, and simple moments of connection, through which much urban sociality is organised.

In part, this involves considering the place of all sorts of frequently overlooked practices and events. Things like friendship, or what Amin and Thrift (2002: 45) call 'light sociality' – 'groups that come together briefly and then disperse again' – or the everyday interactions of bus or shopping queues, to name just a few examples. Alternatively, it also involves thinking about the strange passions that circulate through cities. All sorts of enthusiasms from football, to train spotting, to sun bathing, to angling, to jazz music, to Ronan Keating (the list is quite literally endless), animate people, bring them together, and in the process generate all sorts of forms of connection and belonging that are both less and more than a community. We might also consider the ways that all sorts of non-human actors are involved in the generation of connection, and maybe even participate in community on an equal footing with us (see **materiality**). Think of the role of software in the internet, or in communal gaming environments. Or think of the communion afforded through the headphones of a walkman or MP3 player. In short, then, to think about connection, and beyond community, in this way is to think about all the ways that everyday interactions and everyday materials are bound up in the enchantment of our urban reality (Bennett, 2001).

## KEY POINTS

- Urban theory has long been shaped by the idea that cities are defined by an absence of genuine community.
- Scholars from sociology, anthropology, and geography have shown that community is not 'lost' in modern cities. Rather it has taken on a range of new forms.

- The media plays a central role in the construction of community in the contemporary city. It helps generate a sense of connection among otherwise diverse urban populations.
- The idea of 'community' is not just about inclusion. It can also be employed to exclude those who are seen as different or other to the majority.
- Rather than focus on community, many urban geographers now prefer to consider the dynamics of social connection, or communion.

## FURTHER READING

Claude Fischer (1982) *To Dwell Among Friends* provides a very thorough and highly readable account of the ways community is organised in modern cities. Phil Hubbard's (2006) *The City* provides a very accessible overview of how urban geographers think about community and social connection. *Cities: Reimagining the Urban* by Ash Amin and Nigel Thrift (2002) provides some provocative ideas about how urban geographers might rethink community. See especially Chapters 2 and 4.

*AL*

# 5 Sites and Practices

# 5.1 CONSUMPTION

Consumption has emerged as one of the key concepts in the social sciences over the last two decades, spawning a vast literature and research agenda, and crossing disciplinary boundaries from sociology to economics, psychology to anthropology. Geographers have contributed to this debate in two main ways. The first is through the articulation of processes by which consumption is organised spatially. While original urban settlements had a tight relationship to their surrounding fields and orchards for foodstuffs, the growing complexity of urban societies has long severed this relationship. In other words, 'Demand resulting from agglomeration and population density is not met through local supply alone. Local consumption does not return the city as a closed economic circuit' (Amin and Thrift, 2002: 70). Geographers have become interested in the explanation of commodity chains, for example, or how items gain value and are materially distributed from raw material through to consumed product (e.g., Hughes and Reimer, 2004). The second main approach is through an examination of the sites of consumption in cities. Here, we limit ourselves to a discussion of the latter, given the importance of cities as important demand generators for goods and services, and also as sites for consumption as an *experience*.

Spaces of consumption can be found almost everywhere in the city. Ironically, while the home is probably the most important site of consumption, there has been an overriding focus on *public* sites. This is evident in the myriad of billboards advertising everything from fast food to luxury designer clothing you may see as you travel along urban streets (Cronin, 2006); in theme parks, such as Disneyland and Universal Studios (Hannigan, 1998); in foodscapes, from the hawker centres of Singapore, the street cafes of Paris, and the global food chains in American cities such as Planet Hollywood and Starbucks (Smith, 1996; Klein, 2000); in chain clothing stores or fashion boutiques (Crewe, 2001; Breward and Gilbert, 2006); and in the promotion of particular tourist sites, and their associated consumption infrastructures from airports to hotels (McNeill, 2008b).

However, there has been a tendency to see such spaces as 'all new'; neglecting deeper, historical accounts of the emergence of different consumption forms in different national societies. According to Glennie and Thrift (1992), 'modern' consumption patterns can be traced to the development of an imperial system of trading from the late seventeenth century, where *novelty* became an important aspect of commodity choice, evidenced in 'new types of foreign goods: tea, coffee, sugar, chocolate, tobacco, new fabrics, and various ceramic, wood and metal items' (p. 429). In turn, these new goods fed into new social practices, such as drinking tea and coffee, and smoking. Thus it is important to see consumption as being an urbanised practice, one that has grown and changed over several centuries, and one that is related to changing social practices, hobbies and skills. This has gradually evolved into a growing reflexivity among consumers: in other words, 'the ability of subjects to reflect upon the social conditions of their existence' (p. 436). Contemporary societies are marked out by what some have called 'consumer citizenship', where individuals exercise their rights to be protected from, investigate the source of, and even return, commodities that do not fit within a subject's embodied knowledge about how things *should* be.

# The commodity self

We can enter the world of consumption through the realization that mass-produced products are not only objects having exchange value and use value in the sense of food, clothing, shelter, and entertainment. They are also symbols conveying meaning. An important device imparting meaning to products is advertising, which can be thought of as the language of consumption. It presents commodities as devices enabling individuals to create their own contexts, their own worlds, however ephemeral or enduring they may be. (Sack, 1988: 643)

For a number of theorists, the self is expressed through the purchasing of commodities. Through advertising, consumers are encouraged 'to think of commodities as central means through which to convey their personalities.' (Sturken and Cartwright, 2001: 198). The 'commodity self' expresses the idea that 'our selves, indeed our subjectivities, are mediated and constructed in part through our consumption and use of commodities' (Sturken and Cartwright, 2001: 198). The idea of shopping as a social activity is a key part of this argument, 'a form of leisure and pleasure and as a form of therapy ... [where] commodities fulfil emotional needs' (Sturken and Cartwright, 2001: 197). This has been attributed to 'a gradual shift from a nineteenth-century (and earlier) model founded on ideas of *character* to performative models of *self* in which appearance and bodily

**Figure 5.1.1**   'The packaging and display of commodities is a key element in their consumption'

presentation "expresses" the self' (Glennie and Thrift, 1992: 437, summarising Featherstone, 1991).

A key mechanism for relating to consumers is through the practice of advertising. The use of space and the choice of place are both crucial elements in advertising and consumption. Billboards, postcards, television screens, posters are all located somewhere. Advertising in cities adds to the 'visual mix' (Cronin, 2006: 615) of the urban landscape (See **photography**) and according to some commentators 'create the very fabric of place' (Cronin, 2006: 616). It should be said here that the meaning of commodities, i.e., modes and practices of production, distribution, waste and pollution are often hidden from the consumer in advertising practices.

The visuality of advertising has also been extended to other spaces of consumption, such as the shopping mall, the theme park and heritage sites. Spectacular leisure and consumption spaces such as Disneyland are centred around the consumption of things like food, souvenirs and 'rides'. As well as objects, these sites are about selling 'the experience' and the brand. These sites of consumption are often designed to 'create a placeless environment' (Mansvelt, 2005: 59), where consumers can leave reality behind and enter a carefully designed landscape of fantasy (Hannigan, 1998). Other spaces of consumption include 'cultural centers, cinema complexes, sports stadia, shopping malls, restaurants, and art galleries'. As Mullins et al. (1999:45) claim, there is a 'sociospatial relationship that exists between where households reside and those parts of their metropolitan areas that have been especially built or redeveloped to encourage people to visit so that they can buy and consume some of the many goods and services on sale there'.

Indeed, the design of the buildings often revolves around the consumable fantasy experience, seen most markedly in the likes of Universal CityWalk, Disneyland and Las Vegas. Architecture critic Ada Louise Huxtable (1997) names architectural structures built specifically as entertainment spaces as 'Architainment'. These places are, of course, places to make money, but they are also stages of performance for an interactive consumer. This analysis is important in understanding that specific spaces in the American built environment and landscape are created specifically with the idea of representing popular images of and for an 'audience'. Here, the landscape *becomes* the entertainment.

163

# Retailing as urban experience

An important body of literature has developed around geographies of retailing. Wrigley and Lowe's *Reading Retail* (2002) pins down the two strands of consumption geographies – the spatially stretched networks of retail supply, the sourcing decisions of major supermarkets, the 'inconstant geography and spatial switching of retail capital', along with four of the key sites of retail consumption covered by geographers – the street, the store, the mall and the home. Attention has increasingly focused on the environmental psychology of shopping, most obviously in Goss's studies of the West Edmonton Mall and the Mall of America (1993, 1999), and replicated more convincingly in the retail consultant Paco Underhill's popular books *Why we Buy* (2000) and *The Call of the Mall* (2004). More broadly, and despite an over-association of malls with North America, there is growing interest in the place specificity of malls. *The Harvard Guide to Shopping* (Chung, 2002) supplies a treasure trove of contributions exploring retail cultures worldwide. Erkip's (2003)

**Figure 5.1.2**   'Mercato Mall, Dubai, is designed to mimic an Italian marketplace'

Ankara-based study reveals a complex story of social modernity, simultaneously liberating some women from established social roles, while acting to segregate on class grounds. With the growth of online shopping leading to a new geography of 'bricks and clicks', there has been some concern that the act of shopping would be completely reconfigured, but at present this would appear to be unfounded. There is also an interesting set of literatures which explore 'marginal spaces' of consumption, such as car-boot sales and charity shops, both of which highlight the importance of the life history of commodities (e g., Crewe and Gregson, 1998).

## Foodscapes

Of course, the commonsense meaning of consumption is often taken to be that of eating and drinking, and charting the spatiality of this process has been of growing significance for geographers. In *Consuming Geographies: We Are Where We Eat* (1997), David Bell and Gill

Valentine outline a number of scales where the consumption of food can be conceptualised. This runs from the body, to the home, the city, the region, the nation, and global cuisines. They argue that eating out is a 'container of many social and cultural practices, norms and codes' (p. 125), from the formality of elite restaurants, through to the social rituals of café culture (Laurier and Philo, 2006a, 2006b).

Another way of conceptualising this is through the concept of the 'culinary journey'. As Jean Duruz (2005) has illustrated with a study of two shopping streets in Sydney and London, food allows broader identities to be 'staged'. The streets she chooses – King Street in Sydney's Newtown, and Green Lanes in London's Haringey – are both places of travel, but in an elusive sense. Here, '[the] everyday practices of food shopping, cooking, and eating associated with these streets provide spaces for negotiating meanings of home, ethnicity and belonging.' (p. 51). For migrants, the practice of shopping in these streets is an important enactment of their diasporic identities – in other words, recreating links with homelands through the food bought and consumed, even decades after their departure from that territory. Alternatively, such 'ethnic' foodscapes may provide a means of travelling and experiencing other cultures without physically leaving one's own city. However, for Duruz, it also becomes a way in which such landscapes are evocative of complex forms of intercultural interaction, where food consumption is woven into the lives of long-time 'Anglo-Celtic' Australians or English.

165

With all this focus on consumption, it is unsurprising that there is an increasing concern about how some people are overdoing it. The growing rates of obesity in western societies, particularly among the young, poses a major social challenge. As a UK government committee noted that 'paradoxically, the phenomenal increase in weight comes at a time when there is an apparent obsession with personal appearance. There are more gyms than ever, more options presented as "healthy eating", and the Atkins diet dominates the best seller charts' (House of Commons Health Committee, 2004:). Health professionals are increasingly concerned with the idea of the obesogenic city, in other words, an urban environment seemingly designed to induce the fattening of humans. There are several causes of this: first, there is a proliferation of cheap fast food chains offering large portions of food for very cheap prices; second, there is a lack of attention paid to creating pedestrian or cyclist environments, or to creating structured fitness regimes in everyday urban life; third, there is the problem of 'food deserts', linked to the economic geography of retail location. For example, Whelan et al. (2002) studied economic and physical reasons for geographical inaccessibility to fresh food in urban areas.

Economically, low income families have limited choices in the range of fresh food options as retailers see marginalised places as being uneconomic locations; physically, major supermarkets may be poorly connected by public transport and be poorly designed to allow non-vehicular pedestrian access. There is also a fourth element of interest to urban geographers and sociologists: the visual seduction of consumers in the urban landscape, through brands, icons and adverts. The most notable of these studies is the 'McDonaldization' thesis elaborated by Ritzer (1997), and Naomi Klein's anti-branding manifesto *No Logo* (2000).

What does this mean for urban theory? A number of scholars have approached the relationship of obesity and fatness to the city (Sui, 2003; Longhurst, 2005). Marvin and Medd (2006) use metaphors of 'metabolism' – normally associated with the human body – and transpose them on to understanding how the city deals with fat: 'the contingencies and metabolisms of fat in bodies (as individuals), cities (as a collective site of action), and sewers (as infrastructure) ... highlight a multiplicity of urban metabolisms' (p. 313). They argue that while fat is often associated with mobility (the global production and distribution of meat and dairy products by agribusiness, particularly), it is also key to unlocking *immobility*. Major cities in the US have now been ranked in terms of the environmental factors – from junk food outlets per capita through to their 'keep fit' infrastructures of gyms and recreation areas – which affect the fitness of their urban populations. They describe Philadelphia's attempt to lose its status as fattest city in the US, with an active mayor urging a programme of 'collective weight loss', through encouraging exercise time and targeting restaurant provision of fatty foods. However, there is also an infrastructural challenge, with sewers increasingly clogged up by fatty discharges from restaurants. And while the fat city competition operated at a very visual level, 'making visible the problem in sewers continues to be a challenge for local authorities and utilities charged with sewer maintenance' (Marvin and Medd, 2006: 318).

Such 'fat cities' raise a broader issue concerning the nature of resource use in cities. As Wolch argues, geographers

166

have just begun to confront urban consumption in terms of the geographies of urban footprints, product-specific ecological damage, and transpecies justice, or to articulate a geographical research agenda ... We need sustainability-oriented studies, using a variety of analytic approaches, of alternative urban food regimes, local currency systems, shared-car programs, off-the-grid housing developments, 'social marketing' campaigns to cut consumption, and sociotechnologies such as Craig's List ... that promote reuse. (2007: 377–8)

This is an important and challenging research direction for urban research on consumption, and one that will assume increasing importance in the coming years.

## KEY POINTS

- Consumption has been a major interest in the social sciences in recent times. Geographers are interested in the sites and practices where consumption takes place, how particular sites may encourage or facilitate consumption, changing historical and inter-cultural perspectives on consumption behaviours, and how particular social relationships may be located in particular sites.
- Consumer landscapes are sometimes 'disguised' to frame the user as an 'audience member' rather than an economically targeted consumer. Practices of advertising encourages consumers to exercise their 'self' through their choice of commodities.
- Eating in public spaces can be considered a 'democratising experience' as opposed to an experience reserved for the wealthy. Food consumption is a complex mode of sociality, from commercial cultures which encourage overconsumption, to the informal intercultural exchanges that take place in cities.

167

## FURTHER READING

There is a huge literature concerning the sociology, geography and anthropology of consumption. Some good starting points for its contribution to urban experience include Chung (2002) and Wrigley and Lowe (2002), who both provide comprehensive discussions of the spatiality of retail, including its impact on the urban landscape; Hannigan's *Fantasy City* (1998), which summarises the American urban experience of spectacularly designed spaces; and Jackson et al.'s *Commercial Cultures* (2000), an edited collection that brings out the historical and economic practices that constitute contemporary consumption.

*KM/DMcN*

# 5.2 MEDIA

Recent years have seen the expansion of different formats of media. The 'free to air' programming of national television stations has been challenged by various forms of cable and digital television. Talk-back radio allows audiences to air their views, usually on controversial topics. The development of the internet has allowed the proliferation of online stores, celebrity fan sites and news channels. Mobile phone technology has allowed for increased consumption opportunities via downloads. Even traditional print media, such as the newspaper, now work between paper and electronic formats. Indeed, these types of media overlap with one another on a daily basis, often *converging* where events captured in one media are reported and debated in another media. These media technologies are prevalent in the everyday routines of public life, whether it be glancing at advertising billboards or newspaper stands, using mobile phones, or listening to car radios while driving.

This entry considers how media are important both within fixed spaces of the city, but also within a relational understanding of urbanity. There is a growing interest in 'the links that media objects forge *between* spaces' (Couldry and McCarthy, 2004: 2). What occupies these 'spaces between' are the flows of communication and images, distributed between producers and receivers, or consumers (who we can think of as audience members). These flows are most important, as they *connect* otherwise distant places to each other. So, it is important to consider the nature of the production, distribution and consumption of media culture, and the media *outlets* which distribute and broadcast the information, such as television and radio stations, news offices and agencies, photography and agencies. This has obvious relevance to urban geography. For example it prompts questions such as where is the media made? Where is it broadcast from, and to? How is it distributed? Where are audiences located? How does media flow and converge? In short, what is its geography?

## Geographies of images

As the entry on **photography** suggests, pictures of things are important constituents of modern cities. For example, as Nigel Thrift argues:

... images are a key element of space because it is so often through them that we register the spaces around us and imagine how they might turn up in the future. The point is even more important because increasingly we live in a world in which pictures of things like news events can be as or more important than the things themselves, or can be a large part of how a thing is constituted (as in the case of a brand or a media celebrity). (Thrift, 2003: 100)

Thrift's 'geography of images' suggests a focus on the circulation, travelling images, flows and distribution processes of the media. These currents are not easy to define in straightforward terms, and many scholars have opted for using metaphors to explain the interconnectivity and multi-directional flows of media practices and images in everyday life. Paul Virilio uses the term 'vectors' to capture how images travel, a theme developed by McKenzie Wark in *Virtual Geography*:

Virilio employs it to mean any trajectory along which bodies, information, or warheads can potentially pass. The satellite technology used to beam images from Iraq to America and on to London can be thought of as a vector. Media vectors have fixed properties, like the length of a line in the geometric concept of vector. Yet that vector has no necessary position: it can link any two points together. (1994: 11)

169

Other metaphors have also been used to try to capture these flows: Couldry and McCarthy use the idea of the 'grid' (2004: 1–18) to explain these connections; Charles Acland (2003) speaks of the 'traffic' in distributed images; Mitchell Schwarzer uses the term 'Zoomscape' (2004) to help explain the in/tangibility of the moving image in urban perception; while globalisation theorist Arjun Appadurai (1990: 295–310) employs a metaphor of the 'scape'. These 'scapes' are particularly helpful in charting the perspectival nature of 'new global cultural economy', which revolves around flows of people, media images, technology, finance capital and ideas. This new cultural landscape 'has to be understood as a complex, overlapping, disjunctive order, which cannot any longer be understood in terms of existing centre-periphery models' (Appadurai, 1990: 296). In other words, events and images from geographically distant places can be beamed instantly onto screens around the world.

The **infrastructure** which grounds these media outlets is an important factor to consider. Cable for internet, TV and the tele-phone is threaded through the urban landscape – both visibly and invisibly – holding together a vast network of communications that are vital to the city's smooth and effective working. As Stephen Graham notes, the ubiquity and invisibility of such infrastructure in urban space is paramount, yet mostly unknown to the general public, who do not think about these networks 'beyond the power point, beyond the telephone, beyond the car ignition key, or beyond the water tap or toilet' (Graham, 2000: 184).

Of course, the media is not only on the street, but also in the home, and moves between the two with an increasing amount of ease. Like the smart office building in Mike Davis' scanscape (see **surveillance**), the home can now be wired up to enhance domestic life. As Fiona Allon notes:

> Equipped with 'smart' networked infrastructures, the houses wired by (smart) systems are increasingly integrated into global spaces of exchange and communication where they can tap into ever-diversifying circuits of media and information. The smart house is also a site that negotiates the need for extension with the need for emplacement, enabling secession from public life at the very same time as networked communications achieve greater levels of integration and interconnection. (2004: 256)

170 As Allon continues, the media thus demonstrates the way in which the home's boundaries are now more fluid and, simultaneously, showcases the reach of digital technologies – which can now be accessed and utilised in a domestic setting (Allon, 2004: 253–5). What is certainly true is that an infrastructure of connection (from the old-fashioned aerial to the newer cable or wi-fi networks) brings television and its content into the home, leading some to argue that the public–private dimension to life is partially eroded.

# Cultural economy and media production

There is now a strong interest in the so-called 'cultural economy', where film, TV, radio, photography, advertising and a host of other industries are major sources of jobs and revenue. Contemporary work ranges from policy analysis and planning (Cunningham, 2004; McGuigan, 2004) to the cultural economy and economic development of cities (Landry and Bianchini, 1995; Hall, 2000; Scott, 2004), and of the importance of media infrastructure (Jenkins, 2004). Perhaps most influentially, Richard Florida's (2002) book *The Rise of the Creative Class* explores the idea that cities which house creative people are economically more successful and, indeed, can be regenerated through the creative activity produced by these people (see **centrality**).

So, as opposed to earlier versions of the 'culture industry' (as popularised by the Frankfurt School in the early post-war period, now translated and reprinted, e.g., Adorno, 2001; Adorno and Horkheimer, 2002), which saw media as a generally negative, dominating force, academic studies of the influence of cultural industries are increasingly upbeat. The city has featured as the stage on which mass media could flourish and serve as a way to encourage the development of popular culture in various ways. One of these ways would be the cultivation of individual personalities, or celebrities, which can be dated to the emergence of mass newspapers (including the first tabloids such as the *New York Post*) in the early twentieth century:

> One of the most remarkable traits of the mass-circulation press was its ability to make ordinary people visible at a time when urban growth appeared to be submerging individuals into an anonymous mass. Illuminated by the press, men and women gained a new kind of conspicuousness and were transformed into public figures ... As the press expanded its coverage to include new areas of social life, the criteria journalists employed to determine newsworthiness expanded as well. (Ponce de Leon, 2002: 48)

**171**

In the same vein as reality TV shows are produced today, here Ponce de Leon (who focuses on the US experience) shows how the urban served as a catalyst for the mass media to create public personalities from people in the street. Through a focus on 'human interest' stories, such newspapers reported on crime scenes, city hall, high society gossip, court news, theatre reviews and commercial establishments (see also Park, 1923; Lindner, 1996). They also allowed for the expansion of advertising, which tied newspapers into local 'growth coalitions' (Logan and Molotch, 1987). This also led to the development of celebrity culture:

> By the mid-nineteenth century anybody involved or associated with a news story or an institution regularly covered by the press was a legitimate candidate for journalistic publicity and transformation into a public figure ... obscure people whose travails, once brought to the attention of the press, became the grist for colourful, often poignant human-interest stories. (Ponce de Leon, 2002: 48)

Of course, the same is true today with many tabloid newspapers, magazines and TV shows showcasing the tragic or celebratory stories of everyday people. Furthermore, blogs (such as blogspot.com), personal image banks (such as flickr.com) and various email providers allows the audience to become more than a passive viewer or listener, and can facilitate the creation of media content on a globally accessible internet-basis.

## Media, audiences and place

It has been suggested that the new accessibility of media technology has an impact on social behaviour. An important contribution here is *No Sense of Place* (1985), in which Joshua Meyrowitz argues that technology reorganises social settings, and that this will ultimately reduce the significance of geographical location:

> One can now be an audience to a social performance without being physically present; one can communicate 'directly' with others without meeting in the same place. As a result, the physical structures that once divided society into many distinct *spatial* settings for interaction have been greatly reduced in social significance. (Meyrowitz, 1985: vii)

172

Because communication has changed so radically in recent years, social relations have changed alongside it, to form a 'new social landscape', where contemporary media informs lifestyle. Some argue that these media processes have facilitated the collapse of physical distance and created a one-way flow of information, which has resulted in a global homogenisation of cultural identities. However, Couldry and McCarthy (2004) highlight the 'scale-effects' of media processes, that is, the difference between local and global production, consumption and distribution, especially in terms of how these effects are understood in specific places:

> The emerging picture is not, then, the collapse of place – indeed, our reasons for travelling to distant places to which media connect us have increased, not diminished – but instead the more subtle integration of our interactions with other places and agents into the flow of our everyday experience. (Couldry and McCarthy, 2004: 9)

The audience is a crucial factor in these imaginings of mediated space. As consumers of the image, audience members can even travel to the image (as in Couldry's (2000) study of visitors to the *Coronation Street* set), or have it come to them (in the case of cinema multiplexes). It is important to recognise audiences as a key factor in the way in which media are located, and from where and to, and how mediated information travels. Internet audiences are different from TV audiences and from radio and newspaper audiences. Furthermore, different types of media conceptualises its audience according to its own format. The same reader of a local newspaper will look at the same information differently to how s/he sees it on the internet. Indeed, once web-bound, this information can be seen almost anywhere perhaps changing the nature of the audience member – from passive reader to avid searcher or creator. As Karen Ross and Virginia Nightingale suggests on the emergence of technology and interactivity:

> The development of the world wide web and its gradual journey to becoming a mass medium requires us to rethink both the audience and the medium. The concomitant development of cyber cafes, telecentres and community-based facilities in even the most remote parts of the world, means that the *potential* of the medium to attract a mass audience is considerable. (2003: 147)

Due to this technology, mediated information and images are available in places and at times they had never been before. It allows local access of global news, and vice versa.

# Travelling media: from Hollywood to the multiplex

While the idea of Appadurai's 'scapes' are important in acknowledging the influence images have in constructing different media narratives, we must also consider the physical infrastructure – that is, the materiality and political economy – that allows images to be projected and consumed. For example, in *Screen Traffic* (2003), Charles Acland explores the distribution and marketing process that allows Hollywood films to extend their global reach. He analyses distribution statistics, the construction of multiplexes, the success rate of theatre chains in the United States, cinema-going frequency, the rise of the shopping mall as an entertainment space and the idea of the cinema appealing to audiences as a place to experience a form of 'travel'. As he explains:

> Here is a set of technologies, practices, and shared engagements that cannot be found anywhere else. A trip to the cinema, the passage through the lobby and the consumption of food, drink and games, is a part of the

preparation for the screening – preparation for that filmic twilight – like an urban and architectural trailer for movie watching. Production, distribution, and exhibition, as the broad divisions of the film industry apparatus, present a narrative path for the film commodity as it moves from production to consumption. Clearly, the mechanics of the film business involve not only the marketing of movies but also their delivery to an audience, the gathering up of that audience, and the provision of a site for the film encounter ... Delivery, distribution, and exhibition of film to some segment of the population might be understood best as shaping that segment ... This shaping is not the province of textual conventions alone but of spatial and temporal ones. The 'where' and 'when' of film are crucial components in the formation of audiences, whether imagined as the product of local practices or as manifestations of international popular taste. (Acland, 2003: 229–30)

This analysis is critical to grasping the way media flows operate on a practical level, and is the launching pad for a wider debate about how information and images are circulated once the work at the 'ground level' (e.g., sites of production, distribution and consumption) is completed.

However, when a Hollywood film or TV show is made, how important are place-specific factors in the context of its production? For example, a movie may have been filmed in Toronto, but be set in New York. As production crews are less expensive to hire in Toronto than in New York, this makes economic sense. The film's value is also tied to people who feature in it. So, a big Hollywood star is more of an assurance to the film's financers than an unknown actor or actress. However, a star's salary will be more costly to a studio than that of an unknown. Indeed, they are an important part of the process at every stage in production, distribution and consumption. In the case of films, they are often made with a distinct actor or actress in mind. When distributed, the star will often travel to promote the project, and in preparation for its consumption, it is important that the films have a widespread fan base, which is often achieved through more informal media content such as fan websites.

While Hollywood is the undisputed centre of global film production, other media capitals are emerging as formidable forces within the entertainment and news industries. As Michael Curtin explains, these capitals, such as Hong Kong

represent centres of media activity that have specific logics of their own; ones that do not necessarily correspond to the geography, interests or policies of particular nation-states. For example, Hong Kong television is produced and consumed in Taipei, Beijing, Amsterdam, Vancouver, and

Kuala Lumpur. The central node of all this activity is Hong Kong, but the logistics that motivate the development are not primarily governed by the interests of the Chinese state, or even the Special Administrative Region (SAR). (Curtin, 2003: 203–4)

So, this notion of trans-national television also *grounds* media outlets – as hubs for communication vectors – in certain cities. In this case, Hong Kong acts as an 'intersection of complex patterns of economic, social and cultural flows ... a nexus or switching point, rather than a container'. (Curtin, 2003: 204).

We can also map the distribution of films onto the urban fabric itself. In his study of the locational and design factors of cinema **consumption** in Leicester, England, Phil Hubbard (2002, 2003) highlights the growth of multiplex cinemas, often located in retail parks or suburban malls far from city centres. These cinema complexes are newly built, designed to provide a wide range of films (most complexes usually offering between 15 and 20 screens), with more comfortable surroundings than traditional inner city cinemas. This has a strong correlation to the trends towards decentralisation within many cities, where entertainment – as well as housing, shopping and schooling – has shifted into new suburban locations. There is also a counterpoint in the decline and rediscovery of **centrality** in cities, with a strong movement towards the sustenance of so-called 'arthouse' cinemas within gentrified neighbourhoods. These reflect differing media audiences for the cinema product. It is also linked to a diversity in audiences which has a strong spatial location within cities.

175

To recap, Couldry and McCarthy remind us that as 'electronic media increasingly saturate our everyday spaces with images of other places and other (imagined or real) orders of space, it is ever more difficult to tell a story of social space without also telling a story of media, and vice versa' (Couldry and McCarthy, 2004: 1). This theme, developed earlier by Couldry in *The Place of Media Power*, opens up questions about the widespread connections 'between media and territory' which need to be traversed:

... there is the special status of spaces featured in media production: the studios, the locations, the sites of witnessing news events, and so on. If this is correct, the media do not simply 'cover' territory, let alone 'collapse' the boundaries between places. Instead they shape and reorganise it, creating new distances – for example, between studio and home – and building new presences, new places of significance. Taking these two points together, the 'media frame' involves a complex, but quite definite division

between two types of place: a dispersed mass of sites where media consumption takes place, and a much more limited number of sites where the
media are produced. (Couldry, 2000: 26)

In this sense, the idea of place can be seen as an important aspect when
talking about media spaces and practices. The uneven geography of
media presides not only at the places where media is made, consumed
and distributed, but also must be sensitive to the flows which link these
places.

## KEY POINTS

- The production, distribution and consumption of media is a central
  practice in the constitution of modern urban life.
- It is important to recognise audiences as a key factor in the way in
  which media are located, and how mediated information travels to
  reach its market of consumers.
- The geography of media includes not only the places where media is
  made, consumed and distributed, but also the nodes (or media capitals) between which they pass.

176

## FURTHER READING

*MediaSpace: Place, Scale and Culture in a Media Age* (Couldry and
McCarthy, 2004) is an edited book which gives a thorough overview of
the spatial impact of new media. Charles Acland's *Screen Traffic:
Movies, Multiplexes, and Global Culture* (2003) gives an engaging explanation of how Hollywood films travel from the studio lot to the local
Cineplex. Lynn Spigel's paper 'Media homes then And now' (2001),
explores the relationship between the media and the home, focusing
on the idea of the 'smart home', where human behaviour is monitored
and enabled by technology.

*KM*

# 5.3 PUBLIC SPACE

Public space has become one of the central battlegrounds within contemporary urban scholarship. Acknowledged by most urban theorists to be a key dimension of urban life, there is no simple definition of just what public space is, or indeed should be. At its most basic, public space is simply that space used in common by the public. Things are complicated, however, by the fact that the term 'public' covers a range of different relationships. The term public can be used – as we have already suggested – as simply a synonym for collective. In a more formal extension of this notion of the public, it can refer to the spaces that are owned and controlled by the state. Here the double meaning of the term is apparent, as while certain kinds of public buildings and spaces like public libraries, museums, parks, Lidos and so forth are in fact open to all, many others have more restricted access. Parliaments, administrative offices, police stations, water reservoirs – to name just a few examples – offer only limited and highly controlled access *to* the general public although they exist *for* the public. In this case the public being referred to is in fact a legal entity, a notion of the public as being conceptually homologous with the state and its citizens. The public-ness of these spaces is not about their use but rather their ownership by the public.

177

Things are complicated further when we consider a range of spaces and institutions that are neither collectively owned nor collectively managed, but which are nonetheless in certain senses 'public'. Public houses, or pubs, are so known as they offer hospitality for the population in general – 'the public' – as opposed to the space of a private club or private house where hospitality is available only on the basis of a tightly defined membership or invitation. This is a concept of the public based on the notion of access to all who wish – and can afford – to use it. It is not based on ownership – most pubs are privately owned and privately operated. A parallel notion of the public is represented in shopping precincts and shopping malls. In shopping precincts the situation is somewhat confused as in most – but by no means all – cases the sidewalks of such precincts are publicly owned. Nonetheless, the interior spaces of the stores and other amenities, as is the case with shopping malls, are not. Legally at least they are quite clearly private property. In practise and in terms of everyday experience they are, however, public. They are open to, and given life by, the public. Lastly, there is further meaning of 'public' that maps only loosely onto public space. This usage refers simply to the fact that something is on 'view' to the general population, and it is common circulation. Again this can have a range of meanings. One seeks publicity or to publicise an event, one publishes a book or article, one puts on one's 'public face', while an actor or performer performs before their public.

## Public space and the public sphere

If the notion of what is public, how something becomes public, and the essence of public-ness, is far from straightforward, until recently urban geography has rarely grappled with this ambiguity. Within urban geography, the conception of public space as publicly owned, open to all, and somewhere where one may legitimately make political claims upon and through – what we might call the 'republican tradition' of public space, because it sees the cultivation of a vibrant urban public as central to democratic life (see Amin and Thrift, 2002) – has dominated most theorising of public space.

This tradition of thinking owes much to two pioneering texts; German philosopher Jürgen Habermas's *The Structural Transformation of the Public Sphere,* originally published in 1962 (1989 in English), and the American sociologist and historian Richard Sennett's (1977) *The Fall of Public Man.* Both these works present a history of the emergence of the notion of modern democratic public from the late seventeenth century onwards. Both also argue that a central factor in this emergence involved the creation of places and institutions autonomous from the rule of the feudal ruling classes. Finally, both also argued that most western democracies had witnessed a precipitous decline in public life in the second half of the twentieth century. In Habermas's argument, newly prominent merchant and commercial classes came to develop a series of public institutions – in the fields of art and literature and in commerce – based around a radical notion of individual equality and the rule of reason. Habermas sees in the bourgeois public sphere with its particular balancing of private, public and state, as offering in its content and – for those included within it – a genuinely egalitarian public realm.

Sennett tells a slightly different story. Focusing more than Habermas on the nature of the interactions that structured urban public life – street life, theatres, drinking in coffee houses, visiting city parks, etc. – Sennett explores the ways in which a range of distinctive mannerisms in the Europe of the eighteenth and early part of the nineteenth century assisted in the emergence of a notion of the city as the embodiment of an inclusive public realm. The corrosion of royal monopoly and declining importance of the court through the eighteenth century went hand-in-hand with a radical democratisation in the use of the 'public amenities' of the early–modern metropolis. Sennett argues a key part of this democratisation involved a sustained search for, and experiment with, ways for organising and ordering this emergent public domain. This involved a strict sense of separation between the notion of what was public and what was private, and it also led to the emergence of the notion of public space as essentially a theatrical space – a space where people interacted through complex rituals of artifice and play.

## Consumption, urban spectacle, urban revanchism and the end of public space

As has already been mentioned, Habermas and Sennett saw these democratic achievements being undermined within contemporary society. For Habermas the source of this decline was a complex combination of the rise of a self-oriented consumerism, the commercialisation of the media, and the media's increasing collusion with the capitalist state. For Sennett the origin of decline was somewhat more subtle: the emergence of new notions of self-authenticity, in combination with changes within the mass media and the ways people thought about exposure in public, undermined the institutions that had earlier allowed a vigorous public life to flourish.

Neither Habermas or Sennett are nostalgic thinkers. They did not wish to write histories of decline. But their arguments – for all their subtlety and nuance – have provided a basic narrative template for many subsequent studies of public space, and this template centres on a notion of some kind of historical decline or erosion of the public-ness of urban public space. Thus, if there is one thing that most urban geographers seem able to agree on it is that the traditional public spaces that animate the public life of our cities are in some way in danger.

Habermas's and Sennett's texts are important (if now often forgotten, or under acknowledged, or even overlooked) texts in urban geography's republican tradition of writing on public space. Nonetheless, Michael Sorkin's (1992) edited collection, *Variations On A Theme Park: The New*

*American City and the End of Public Space*, is the book that actually came to define the trajectory of urban geography's engagement with public space. It, along with Marshall Berman's earlier (1982) *All That is Solid Melts into Air*, reaffirmed, reframed and restated the notion that a city's urban public spaces were a central index of the social and political health of that city. Crucially, however, Sorkin and his collaborators provided an account of the decline – indeed 'the end'! – of public space that placed geographical explanations of public space's decline at its dead centre.

Extending and spatialising many of Habermas's arguments about the corrosive powers of mass consumerism and the hyper-mediatisation of society, Sorokin et al. argued they could trace out the emergence of a new urban paradigm in American cities. On the surface much of this 'new city' (Sorkin, 1992: xii) appears to be enormously positive. Across America, politicians, property developers and middle class communities were re-discovering the pleasures of urban life. Processes such as inner-city gentrification, along with the emergence or revival of festival market places, street markets, farmers markets, hospitality strips, and all sorts of lifestyle-oriented urban regeneration schemes, had transformed and apparently enlivened whole areas of America's larger cities.

But Sorkin et al. argue that this new urbanity, and the public spaces that were at its core, are more simulation than reality. They argued that the new city 'co-opts ... the traditional scenes of urbanity' (Sorkin, 1992: xv), regulating them through CCTV, privatised security firms, and punitive by-laws, into 'mere intersections on a global grid for which time and space are ... obsolete'. That is to say, much of this urban 'revitalisation', for all its 'local' feel, is driven by corporate capital, corporate strategies, whose vision and theatre of action transcends the local. This is a new urbanism with a 'sinister twist.' An urbanism organised around

> an architecture of deception which, in its happy-face familiarity, con-
> stantly distances itself from the most fundamental realities [of the city
> around it]. The architecture of this city is almost purely semiotic, playing the
> game of grafted signification, theme park building. Whether it represents
> generic historicity, or generic modernity, such design is based on the same
> calculus as advertising, the idea of pure imageability. (Sorkin, 1992: xiv)

Thus, while this new urbanism, with its rediscovered urbanity and lively public spaces, appears to be completely different to the 'suburban cities' (p. xii) that have defined America's urban growth since at least

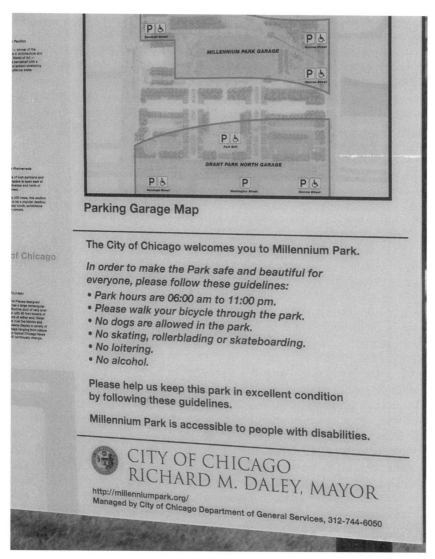

**Parking Garage Map**

The City of Chicago welcomes you to Millennium Park.

*In order to make the Park safe and beautiful for everyone, please follow these guidelines:*
- *Park hours are 06:00 am to 11:00 pm.*
- *Please walk your bicycle through the park.*
- *No dogs are allowed in the park.*
- *No skating, rollerblading or skateboarding.*
- *No loitering.*
- *No alcohol.*

Please help us keep this park in excellent condition by following these guidelines.

Millennium Park is accessible to people with disabilities.

CITY OF CHICAGO
RICHARD M. DALEY, MAYOR
http://millenniumpark.org/
Managed by City of Chicago Department of General Services, 312-744-6050

**Figure 5.3.1** 'Public space regulations in Millennium Park, Chicago'

the late 1940s it is, in fact, simply an extension, or radicalisation, of this amorphous suburbanism. Just as the suburban city gave little weight to public life, so too did this new form of American city.

Now the argument presented in *Variations on a Theme Park* was not without precursors. In a seminal essay, 'Flexible accumulation through urbanization', (1987) David Harvey argued that the widespread rediscovery of powers of public 'spectacle and play' within western cities since the 1970s was the urban face of an emerging 'regime of flexible accumulation' (Harvey, 1989a/1987: 271). Neil Smith (1979, 1987) – who authored a chapter in *Variations* – had been presenting a similar argument in his work on gentrification for over a decade. Mike Davis's (1990) account of the 'fortification' of Los Angeles in *City of Quartz* had become an intellectual bestseller (see also Dear and Wolch, 1987; Leitner, 1990; like Smith, Davis too was a contributor to *Variations*). Nonetheless, *Variations on a Theme Park* crystallised and defined what was to become a remarkably influential approach to understanding what was happening to the public spaces of contemporary cities.

Thus, writers like Jon Goss (1993, 1999), Michael Dear (2000), Edward Soja (1996) and Mark Gottdiener (1997) have analysed the ways in which all sorts of public, and quasi-public, spaces like shopping centres, malls and urban downtowns have become veritable 'urban simulators' (Davis, 1992); creating places that provide a sense of being part of 'a public', while in fact they are carefully calibrated, highly choreographed, machines for generating profit. Don Mitchell (2003) and Neil Smith (1996) have narrated the emergence of a so-called 'urban revanchism' whereby unruly, poverty stricken neighbourhoods in newly-desirable central locations such as the Lower East Side in Manhattan are reclaimed – not infrequently by force, or the threat of force – for redevelopment. Thus, in this account, gentrification, far from being a benign, revitalising, process, in fact often involves the wholesale, and frequently shockingly brutal, 'cleansing' and 'pacification' of inner-city areas to make them 'safe' for middle class residents. And, lastly, a wide range of authors (see Brenner and Theodor, 2002) have tracked the policy circuits through which the public policies – of privatisation, of public/private partnerships, of zero-tolerance policing, of governmental entrepreneurialism, and so forth – that underpin this new urbanism have been sustained and, indeed, reproduced not just in America but in much of the rest of the world also. Where earlier Harvey (1987) had written of a new urban 'regime of flexible accumulation', Neil Brenner and Nik Theodore (2002a,b; Peck, 2005; see also Smith, 2002) now speak of a globally ubiquitous 'neo-liberal urbanism'.

# The end of public space? Really?

Through this collection of accounts, urban geography arrives at an urban landscape where public space has been completely (or nearly completely) eviscerated. This notion of the end of public space has real narrative drama. It also presents a straightforward and easily legible story. With the recognisable villains of self-interested urban elites, large corporations, and a democratically unresponsive neo-liberal state, along with easily defined meta-concepts like spectacle, revanchism, privatisation, neo-liberalism and (for good measure) globalisation, it has a compelling collection of characters and scenes to draw on. That in no small part explains its popularity within urban geography. Yet, it is an argument, or rather a set of arguments, that is in a number of ways quite deeply flawed (see Amin and Thrift, 2002; Latham, 2003b; Bell, 2007; Iveson, 2007).

First, it is flawed because it is based on empirically thin accounts. While it is possible to find examples of all the trends that the decline/end of public space theorists recount, it is equally possible to find an equally convincing set of counter examples. David Ley (1996), for example, provides a rather different account of gentrification to that offered by Smith, suggesting that it has invigorated the public spaces of Canadian cities in all sorts of ways. And, if New York is driven by an urban revanchism, where does a city like Zürich fit into this account, with its green/red coalition city government that places a high value on all sorts of public spaces? (See Bratzel, 1999; Boesch, 2001.) Second, conceptually this literature is based around a narrow interpretation of what 'real' public space is. By emphasising an idealised notion of 'pure' public space (space that is owned, used, and governed by 'the public') *any* movement away from this ideal marks a potential danger to a city's public life (see Iveson, 2007). Third, much of the urgency of this literature comes from a belief in the central importance of genuinely public public spaces for the sustenance of democratic life. Yet, as Ash Amin and Nigel Thrift (2004: 232) note, due to transformations elsewhere this may be less and less the case: 'the public arena and public culture are no longer reducible to the urban (owing to the rise of virtual and distanciated networks of belonging and communication)'. Fourth, and finally, largely because of the failings in points two and three, this literature on the decline/end of public space is ill-equipped to register all sorts of emergent trends that demonstrate the ways public spaces are being collectively reimagined, reinvented and reanimated (Latham, 2003b; Latham and McCormack, 2008).

# Everyday urbanism, new urban publics

How then might urban geography begin to think about public space in ways that move it away from this narrative of decline? The first way to expand human geography's conceptualisation of public space is to broaden its definition of public-ness. Rather than 'seeking a single, all-inclusive public space' (Crawford, 1999: 23; see also Iveson, 2007) urban geographers could look for all the diverse ways that urban spaces are being populated, or re-invented, through popular use. Central to this, is the need for more subtle framings of the relationship between urban public space, urban public culture and commerce. As we have seen, commerce played a central role in Habermas's and Sennett's accounts of the vigorous public spaces and public cultures that defined eighteenth and nineteenth century cities (as indeed it does in Marshall Berman's *All That is Solid Melts into Air*), and urban geography needs to discover ways of acknowledging how commerce can, and does, animate urban culture in all sorts of ways. Writers like Crawford (1999), or feminist urban historians such as Erika Rappaport (2000), or, indeed, Berman (2006) in his most recent book, *On the Town*, offer examples of how this might be done.

184     A second way to extend urban geography's conceptualisation of public space and public-ness, is to explore the new forms of collective life emergent within cities, and consider how these new forms inhabit and animate urban space. Events as diverse as 'fun runs', flash mobs, urban beaches and critical mass bike rides suggest not only novel forms of sociality. They also underline the continuing importance of public spaces, and public gatherings, to city life (see Amin and Thrift, 2002; Latham and McCormack, 2008). A third, and final, way urban geography could expand its notions of public space is to reconsider the **materialities** that constitute public spaces and their public-ness. Take, for example, spaces like roads and motorways. Such spaces tend to be viewed as emptied of their public content, dominated as they are by the private automobile. But if one focuses on the objects and materials that are organised through roads and motorways a rather different picture develops. A whole array of public institutions, public debates, and public anxieties, are focused on regulating, organising, guiding and adjudicating on, the interactions that take place in these spaces – think of speed limits, of laws about drink driving, of road codes, of laws of liability, of arguments about global warming, and so forth (see Sheller and Urry, 2003; Latham, 2004; Latham and McCormack, 2004; Jain, 2004; **mobility**). These publics, to be sure, are rather different, alien even, to the republican idea of public space that

had influenced so much of urban geography's writing on public space. But they do point to all sorts of ways in which cities continue to generate vibrant communal spaces, as well as a whole range of challenging and novel politics of public space.

## KEY POINTS

- Public space refers to spaces that are open to the general population ('the public'). In an ideal sense this would mean public spaces are open to all. In actual fact, few public spaces fit this ideal.
- Urban geographical writing on public space has been dominated by the 'republican tradition'. This views public space as primarily a space available for people to make political claims, and gain public recognition.
- Many urban geographers argue that the public spaces of contemporary cities being eroded by the rise of consumerism, spectacle-isation, of urban space, and a generalised fear of crime and difference.
- This narrative of the decline or even death of public space presents a narrow, and frequently empirically thin, account of trends shaping the public spaces, and public cultures, of contemporary cities.
- Through focusing on the mundane materialities that define urban life it is possible to discover all sorts of interesting and hopeful ways in which contemporary urban public spaces are being re-imagined and re-animated.

185

## FURTHER READING

Loretta Lees' (2004) *The Emancipatory City? Paradoxes and Possibilities* is an excellent collection of essays examining the dynamics of urban public space from a diverse set of perspectives. Kurt Iveson's (2007) *Publics and the City* provides a series of engaging case studies on the construction and contestation of urban public space. Alan Latham's (2003b) 'Urbanity, lifestyle and the politics of the new urban economy' provides an overview and a critique of the narrative of decline that defines much contemporary writing on public space and urban public culture. Bruno Latour and Peter Weibel's (2005) *Making Things Public: Atmospheres of Democracy* explores the ways that notions of what is public, and public-ness, are being transformed.

*AL*

# 5.4 COMMEMORATION

Historical events, memories and people are commemorated in many ways. They can come in the form of statues, buildings, sculptures, gravestones and other such public works. Such commemoration can be made formally by state recognition, such as through government sponsorship of, or the awarding of planning permission to, new monuments. Or it can occur more organically or spontaneously, as seen in the popular pilgrimages to Kensington Palace on the death of Princess Diana, or the popular gatherings around Ground Zero in New York. Spaces where commemoration occurs are usually 'intensely social, animated by an uninterrupted flow of people coming to [for example] mourn, debate, and protest' (Knauer and Walkowitz, 2004: 1). These spaces and objects are endowed with meaning both through large-scale social and political struggle, but also through individual, small-scale acts of remembrance, or even forgetting.

The desire to formally commemorate within the urban landscape is usually carried out through monuments of one sort or other, which are usually constructed with tough materials, usually stone, designed to

186 maximise their functional longevity but also to provide a sense of permanence. Yet they are more than just a lump of stone or metal. Despite their physical solidity, their social significance changes over time, posing a challenge for urban geographers:

> Monuments constructed in the past can become static through time, then get re-energised as they are used ceremonially, as part of a spectacle or commemorative event. They frequently move from a passive space into a dynamic use, then back again ... How this orchestration of a mythic history plays out is reflective of this particular configuration of power relations operative in society at a specific moment in time ... Contemporary cultural geography is thus involved not in documenting and describing the 'traces' left in the landscape or the contemporary nexus of economic and political forces operating to produce those traces, but rather engages in an archaeology of power that is polymorphous and protean, and must be researched in detail using a great variety of sources. (Mitchell, 2003: 446)

So, explaining the significance of monuments in contemporary urban societies is an important challenge to geographers (Johnson, 2002), not least in their centrality to tourist practices. In addition, the location of monuments are important, often sited in parks and squares which are more than just empty spaces, but also spaces of significance, which are given meaning by different, often competing, political elites, historical experts, local citizens' groups, 'opinion formers' such as journalists, academics and museum professionals. And it is also important to consider how certain buildings remain haunted by their one-time inhabitants, be it Hitler's bunker or Elvis's Graceland.

# Memory politics

Certainly, many of the most powerfully charged debates are taking place in areas of recent political conflict. In 1989, mobs targetted Stalin or Lenin statues in many parts of East/Central Europe. In Berlin, the Wall was demolished, in the East, the headquarters of the Stasi (secret police) became an important site of memorialisation. As Karen Till shows in *The New Berlin: Memory, Politics and Place* (2005), the city's urban fabric can be seen as a palimpsest of historical struggle. In other words, layers of historical struggle are imprinted unevenly upon each other in a collection of buildings, memory sites and rubble. To excavate the sites of Gestapo terror and Holocaust commemoration, Till follows through an ethnographic exploration of key places in the city (fieldnotes recording in detail the feeling of these diverse sites), underpinning this with transcripts of interviews conducted with key 'memory-makers' in the city, such as museum curators to historical experts (see also Ladd, 1997).

187

Historians and political geographers are even more interested in how cities act as central points not only in national territories, but also in national narratives, in the continuous telling of histories, expressed in the way history is taught at school, in the way politicians talk about history in their speeches, in displays at museums, and in the urban landscape. In their search to explain how 'nationalistic movements assert the existence of a unified political community – a nation – with a distinct historical narrative, cultural practices, language, ancestry, and territorial home' (Forest and Johnson, 2002: 526), they are interested in how some parts of the urban landscape are more symbolic than others, acting as 'condensation sites': places which allow for the retelling of complex stories in simplified, populist, ways. Such examples include: the museum at Auschwitz, the Neue Wache in Berlin (Till, 2005), the Cenotaph in London, or even street names (Crang and Travlou, 2001; Edensor, 2002). Though monuments may commemorate one particular event, over time they may be re-interpreted in other contexts and become important in terms of a transformative nature of a political event: 'Narratives in public sites ... can be presented in images and displayed, condensed and congealed into monuments, represented in physical spaces or projected through storytelling'. (Knauer and Walkowitz, 2004: 9).

However, commemoration can be problematic for a number of reasons. First, formal, state-sponsored commemorative events are partisan, in that they often emphasise a particular retelling of historical events, and as history tends to be conflictual, it will exclude certain groups. For example, Marston (2002) describes the political contestation over who should be allowed to participate in St Patrick's Day parades in the US. Second, it can 'reawaken ghosts', or stir things up. What to do with Nazi remains? Does vilifying ex-enemies prevent stable democratic settlements? Such was a problem facing post-1989 East and Central Europe, most notably in Berlin (Ladd, 1997). Critics argue that commemoration may seek to 'parcel off' or compartmentalise history, providing closure for what some feel should still be a live debate. Third, it may be elitist, as in the royal statuary found in many cities (representing 'the history of kings and queens' rather than of social history, for example). For instance, a number of critics see World War I commemoration in the UK as excessively triumphalist, representing a glorious national triumph instead of revisiting what some see as the gross misconduct of officer elites, using working class conscripted troops as cannon fodder.

188

Ultimately, this is the result of the contingent relationship between memory and place, which changes over time for various reasons. Indeed, contemporary mass media representations have sometimes altered the perception of monuments. An example can be found in the Hollywood mythologisation of the Scottish nationalist hero William Wallace. Due to the success of the film about Wallace's life, *Braveheart* (1995), starring Mel Gibson, the long-standing monument to Wallace at Stirling in Scotland became the focal point for a restatement of Scottish identity, as well as providing commercial opportunities for the tourist industry and local businesses. This is an important point to consider when thinking about *any* monuments. Part of the experience of monuments is what viewers bring with them: their own experiences and memories, the context in which they are viewing the monument, for example when a tourist joins a bus following a guide, or as an avid researcher makes a pilgrimage. As Tim Edensor writes, monuments should be seen 'in performance':

> Performance is a useful metaphor since it allows us to look at the ways in which identities are enacted and reproduced, informing and (re)constructing a sense of collectivity. The notion of performance also foregrounds identity as dynamic; as always in the process of production ... By extending

the analysis to other theatrical concepts we can further explore the meaningful contexts within which such action takes place ... by conceiving of symbolic sites as *stages*, we can explore *where* identity is dramatised, broadcast, shared and reproduced, how these spaces are shaped to permit particular performances, and how contesting performances orient around both spectacular and everyday sites. (Edensor, 2002: 69)

Thus, a growing number of studies have utilised these theatrical metaphors to describe how monuments are used as 'stages' by political elites (e.g., Atkinson and Cosgrove, 1998).

# Pilgrimage

Monuments are a part of the public history of cities, and are often noted as icons which affect the traffic of tourists and other visitors to the city. For example, as Edensor says of the Taj Mahal:

First and foremost, the Taj, as a globally renowned icon of beauty, has become a symbol of India, yet the building is invested with differently symbolic attributes by different groups. For most foreign tourists, the building is the prime reason for their visit to India, for it has been constructed for the past 150 years as a signifier of the 'exotic East' ... For most Indian visitors, on the other hand, the site represents national pride, most particularly concerning the interweaving of the diverse ethnic, religious and cultural traditions which are believed to signify the Indian 'genius' for cultural synthesis; in addition, the Taj is commonly acknowledged as a place which brings visitors from all over the sub-continent and is a great place to meet fellow Indians ... Contestation over the Taj Mahal is an exemplary instance of the ways in which symbolic sites have a mythical function, in that they can be widely shared as a cultural resource, there is consensus that they are of importance, and yet the values which inhere in this status can be contested. (Edensor, 2002: 46–47)

189

As Edensor notes, iconic monuments such as the Taj Mahal are also sacred, or religious places, for worshippers to come and pay their respect. Graves, or memorials for the dead are also in this category.

In *Elvis after Elvis,* an exploration of the posthumous career of Elvis Presley, Gilbert Rodman demonstrates how the power of the Elvis myth rests in part upon his fans' pilgrimage to Memphis, the plethora and range of merchandise bearing his image, the intellectual and tabloid interest in his life and death, and the culture which has arisen from these elements combined (Rodman, 1996: 1). Rodman suggests that Elvis' death heralded a notion of fame by association in the form of Graceland, a space which became an almost-religious site of pilgrimage for fans of the entertainer, and propelled Elvis' posthumous career and myth-like status. Furthermore, as a physical structure, Graceland can continue to maintain the Elvis legend as long as it stands. As Rodman says:

> Graceland gave Elvis something no other US celebrity of the twentieth century had: a permanent place to call 'home' that was as well known as its celebrity resident ... Graceland altered the shape of his stardom in such a way as to ultimately give rise to his unusual posthumous career. (1996: 99)

To this day Graceland acts as an embodiment of Elvis in a number of ways: spiritually, because he is buried there; physically because it was where he lived, and relatives such as daughter Lisa Marie, and widow Priscilla continue to use parts of the house as their home; and commercially because parts of the house are used as a museum where visitors can take a tour of the mansion. So, commemoration in this case is, as with the Taj Mahal, tied into tourist practices, 'pilgrimages' and travel.

## Media events and 'dark tourism'

Increasingly, the role of monuments in cities has changed. In particular, static commemorations as found in statues are challenged or reconfigured by media events which, as Dayan and Katz note, 'attract and enthral very large audiences, a nation, or several nations' (1992: 5). Such events also frame the cities in which they occur. As Lennon and Foley argue,

> the development of media and communications-driven tourism motivations are a feature of the late twentieth century, essentially because these technologies deliver global events into situations that make them appear to be local (i.e., via television and other news media). These images are then reproduced and reinforced via other media forms (e.g., films and novels). (1999: 46)

These connections are often reified by people flocking to the sites of the event in the days after they have occurred. After 9/11, vast amounts of people from all over the world made a 'pilgrimage' to the remains of the site, dubbed 'Ground Zero'. The site's popularity as a tourist destination took on a life of its own in the months following the tragedy:

> The Ground Zero site itself also became a heavily symbolic setting used for photo-ops during the visits of politicians and dignitaries and a backdrop for

**Figure 5.4.1** 'Commemoration of 9/11 attacks at St Paul's Chapel next to the Ground Zero site'

news broadcasts. Once the site had been secured and the recovery and clean-up efforts were well organized, it quickly became an organized shrine and tourist attraction. (Abrams, Albright and Panofsky, 2004: 208)

This type of tourist attraction rests uneasily between 'patriotic tourism' (Greenberg, 2003: 412), where tourists visit significant national landmarks, buildings, or natural landscapes for historical or nationalistic reasons, and 'dark tourism' (Lennon and Foley, 2000), where visitors travel to sites 'in some way connected to death (e.g., murder sites, death sites, battlefields, cemeteries, mausoleums, churchyards, the former homes of now-dead celebrities)' (Lennon and Foley, 2000: 4).

The viewing tower at Ground Zero had up to 1.8 million visitors in the year following the attacks, and in 2003 – despite the platform's absence – this figure was doubled, drawing 3.6 million visitors. What this meant was that the site had, in effect, become one of Manhattan's top attractions (Greenberg, 2003). Makeshift memorials at Ground Zero itself consisted of international flags, fire and police department t-shirts and badges, teddy bears, candles, wreaths and photographs of the missing on the fence of St Paul's Chapel, opposite the site. Years after the tragedy, these artefacts are displayed in museums, often touring America.

To conclude, interventions in the urban environment are very deliberate attempts to construct a story about the past (and, of course, the present). They are often sources of controversy, especially at times of regime change (as in Berlin and other post-communist cities after 1989). The geography of *location* of monuments is important; as are the aesthetics and form of the statue or monument. And do monuments become *invisible* through over-familiarity? Mitchell articulates a strong case for the particular insights geography can shed on these matters:

> Writers and historians often have a strong abstract awareness of the inter-connections of space, time, memory and recollection, but geographers tend to pursue doggedly, and in far greater detail, the precise ways in which memory becomes embedded in the actual, physical landscape, through the daily habits and movements associated with specific buildings, walkways, monuments, and vistas. (2003: 455)

192

So, while urban history and commemoration is often seen as being primarily a temporal relationship, it also has a powerful spatiality, one that geographers are well-equipped to explore.

## KEY POINTS

- Monuments come in many different forms, and are public spaces for people to mourn, debate and protest. They act as markers of major events and their protagonists.
- Media events challenge the traditional role of monuments in their function as commemorative objects. Media events can also be commemorative acts, linking people to places.
- Tourism to monuments and cities housing monuments is not only commemorative, but also commercial. Souvenirs, photographs and t-shirts are just a few of the physical objects tourists can take away with them.

## FURTHER READING

Ladd's (1997) book *The Ghosts of Berlin* is an excellent overview of the commemoration debates concerning that city's legacy of several controversial regimes, including its role in the Nazi era, and the showcase for Stalinist East Germany (see also Till, 2005). The collection edited by Walkowitz and Knauer (2004) provides a geographically diverse range of case studies and examples which illustrate the range and depth of debates concering political commemoration.

*KM/DMcN*

# BIBLIOGRAPHY

Abrams, C.B., Albright K. and Panofsky, A. (2004) 'Contesting the New York Community: From Liminality to the "New Normal" in the Wake of September 11', *City and Community*, 3(3): 189–220.

Acland, C.R. (2003) *Screen Traffic: Movies, Multiplexes, and Global Culture*, Durham, NC and London: Duke University Press.

Adey, P. (2007) Surveillance at the airport: surveilling mobility/mobilising surveillance, *Environment and Planning A*, 36: 1365–1380.

Adorno, T. (2001) *The Culture Industry*, London: Routledge.

Adorno, T. and Horkheimer, M. (2002) *Dialectic of Enlightenment*, Stanford, CA: Stanford University Press.

Agamben, G. (1998) *Homo Sacer: Sovereign Power and Bare Life*, Stanford, CA: Stanford University Press.

Albrecht, J. (2007) *Key Concepts and Techniques in GIS*, London: Sage.

Allen, C. (2007) 'Of urban entrepreneurs or 24-hour party people? City-centre living in Manchester, England', *Environment and Planning A*, 39: 666–683.

Allen, J., Cochrane, A. and Massey D. (1998) *Rethinking the Region: Spaces of Neo-Liberalism*, London: Routledge.

Allen, J., Massey, D. and Pryke, M. (eds) (1999) *Unsettling Cities*, London: Routledge.

Allen, S. (1999) *Points + Lines: Diagrams and Projects for the City*, New York: Princeton Architectural Press.

Allon, F. (2004) 'An ontology of everyday control: space, media flows and "smart living" in the absolute present', in N. Couldry and A. McCarthy (eds), *MediaSpace: Place, Scale and Culture in a Media Age*, London: Routledge.

Amin, A. (2002) 'Spatialities of globalisation', *Environment and Planning A*, 34: 385–399.

Amin, A. (2007) 'Re-thinking the urban social', *City*, 11(1): 100–114.

Amin, A. and Graham, S. (1997) 'The Ordinary City', *Transactions of the Institute of British Geographers*, NS, 22(4): 411–429.

Amin, A. and Thrift, N. (2002) *Cities: Reimagining the Urban*, Cambridge: Polity.

Amin, A. and Thrift, N. (2004) 'The "emancipatory" city?' in L. Lees, (ed.) *The Emancipatory City: Paradoxes and Possibilities*, London: Sage. pp. 231–235.

Anderson, B. (1983) *Imagined Communities*, London: Verso.

Anderson, K.J. (1988) 'Cultural hegemony and the race-definition process in Chinatown, Vancouver: 1880–1980', *Environment and Planning D: Society and Space*, 6(2): 127–149.

Anderson, K.J. (1991) *Vancouver's Chinatown: Racial Discourse in Canada, 1875–1980*, Montreal: McGill-Queen's University Press.

Anderson, K.J. (1995) 'Culture and nature and the Adelaide zoo: at the frontiers of "human" geography', *Transactions of the Institute of British Geographers*, 20(3): 275–294.

Anderson, K.J. (1998) 'A walk on the wild side: a critical geography of domestication", *Progress in Human Geography*, 21(4): 463–485.

# Bibliography

Anderson, K.J. (1999) 'Introduction', in F. Gale and K. Anderson (eds), *Cultural Geographies,* second edition, Melbourne: Addison Wesley Longman. pp. 1–21.

Anderson, K.J. and Jacobs, J.M. (1999) 'Geographies of publicity and privacy: Residential activism in Sydney in the 1970s', *Environment and Planning A,* 31(6): 1017–1030.

Andrejevic, M. (2004) *Reality TV: The Work of Being Watched,* New York: Rowman and Littlefield.

Appadurai, A. (ed.) (1986) *The Social Life of Things: Commodities in Cultural Perspective,* Cambridge: Cambridge University Press.

Appadurai, A. (1990) 'Disjuncture and difference in the global cultural economy', *Theory, Culture and Society,* 7: 295–310.

Appadurai, A. (1996) *Modernity at Large: Cultural Dimensions of Globalization,* Minneapolis: University of Minnesota.

Atkinson, D. (1998) 'Totalitarianism and the street in fascist Rome', in N. Fyfe (ed.), *Images of the Street: Planning, Identity and Control in Public Space,* London: Routledge. pp. 13–30.

Atkinson, D. (2007) 'Kitsch geographies and the everyday spaces of social memory', *Environment and Planning A,* 39(3): 521–540.

Atkinson, D. and Cosgrove, D. (1998) 'Urban rhetoric and embodied identities: city, nation and empire at the Vittorio Emmanuele II monument in Rome, 1870–1945', *Annals of the Association of American Geographer,* 88(1): 28–49.

Attfield, J. (2000) *Wild Things: the Medical Culture of Everyday Life,* Oxford: Berg.

Augé, M. (1995) *Non-places: Introduction to an Anthropology of Supermodernity,* London: Verso.

Balshaw, M. and Kennedy, L. (eds) (2000) *Urban Space and Representation,* London: Pluto Press.

Barber, S. (2002) *Projected Cities: Cinema and Urban Space,* London: Reaktion Books.

Barley, N. (ed.) (2000) *Breathing Cities: The Architecture of Movement,* Berlin: Birkhäuser.

Barnes, T.J. (2003) 'The place of locational analysis: a selective and interpretive history', *Progress in Human Geography,* 27: 69–95.

Barnes, T.J. (2005) 'The 1990s show: culture leaves the farm and hits the streets', in B.J.L. Berry and J.O. Wheeler (eds), *Urban Geography in America, 1950–2000: Paradigms and Personalities,* New York: Routledge. pp. 311–326.

Barnett, C. (2003) 'A critique of the cultural turn', in J.S. Duncan, N.C. Johnson and R.H. Schein (eds), *A Companion to Cultural Geography,* Oxford: Blackwell. pp. 38–48.

Barthes, R. (1972) *Mythologies,* London: Vintage.

Batty, M. (2005) *Cities and Complexity: Understanding Cities with Cellular Automata, Agent-based Models, and Fractals,* Cambridge, MA: MIT Press.

Batty, M. (2006) 'Digital cornucopias: changing conceptions of the virtual city', *Environment and Planning B: Planning and Design,* 33: 799–802.

Batty, M. and Smith, A. (2002) 'Virtuality and cities: Definitions, geographies, Designs', in P.F. Fisher and D.B. Unwin (eds), *Virtual Reality in Geography,* London: Taylor and Francis. pp. 270–291.

Baudrillard, J. (1988) *America,* London: Verso.

# Bibliography

Bauman, Z. (2001) *Community*, Cambridge: Polity.

Beauregard, R.A. (1993) *Voices of Decline: The Postwar Fate of US Cities*, Cambridge, MA: Blackwell.

Beaverstock, J.V., Smith, R.G. and Taylor, P.J. (2000) 'World-city network: a new meta-geography?', *Annals of the Association of American Geographers*, 90(1): 123–134.

Beaverstock, J.V. (2002) 'Transnational elites in global cities: British expatriates in Singapore's financial district', *Geoforum*, 33(4): 525–538.

Beaverstock, J.V. (2005) 'Transnational elites in the city: British highly-skilled inter-company transferees in New York City's financial district', *Journal of Ethnic and Migration Studies* 21(2): 245–268.

Beckman, J. (2001) 'Automobility – a social problem and theoretical concept', *Environment and Planning D: Society and Space*, 19: 593–607.

Bell, C. and Newby, H. (1972) *Community Studies: An Introduction to the Sociology of the Local Community*, London: Harper Collins.

Bell, D. (2007) 'The hospitable city: social relations in commercial settings', *Progress in Human Geography*, 31(1): 7–22.

Bell, D. and Valentine, G. (eds) (1995) *Mapping Desire: Geographies of Sexuality*, London: Routledge.

Bell, D. and Valentine, G. (1997) *Consuming Geographies: We Are Where We Eat*, London: Routledge.

Bell, D. and Jayne, M. (eds) (2006) *Small Cities: Urban Experience Beyond the Metropolis*, Abingdon: Routledge.

Benjamin, W. (1999) *Illuminations*, London: Pimlico.

Bennett, J. (2001) *The Enchantment of Modern Life: Attachments, Crossings, Ethics*, Princeton, NJ: Princeton University Press.

Berman, M. (1982) *All That is Solid Melts into Air*, London: Verso.

Berman, M. (2006) *On the Town: One Hundred Years of Spectacle in Times Square*, New York: Random House.

Berry, B.J.L. (1973) *The Human Consequences of Urbanisation: Divergent Paths in the Urban Experience in the 20th Century*, London: Macmillan.

Berry, B.J.L. and Wheeler, J.O. (2005) *Urban Geography in America, 1950–2000*, New York: Routledge.

Bialasiewicz, L. (2006) 'Geographies of production and the contexts of politics: dislocation and new ecologies of fear in the Veneto *città diffusa*', *Environment and Planning D: Society and Space*, 24: 41–67.

Bingham, N. (1996) 'Object-ions: from technological determinism towards geographies of relations' *Environment and Planning D: Society and Space*, 14: 635–657.

Biressi, A. and Nunn, H. (2003) 'Video justice: crimes of violence in social/media space', *Space and Culture*, 6: 276–291.

Bishop, P. (2002) 'Gathering the land: the Alice Springs to Darwin rail corridor', *Environment and Planning D: Society and Space*, 20: 295–317.

*Bladerunner* (1980) Directed by Ridley Scott.

Blunt, A. and Dowling, R. (2006) *Home*, London: Routledge.

Boesch, H. (2001) *Die sinnliche Stadt: Essays zur modernen Urbanistik*, Zürich: Nagel & Kimche.

Borden, I. (2001) *Skateboarding, Space and the City: Architecture and the Body*, Oxford: Berg.

Bourdieu, P. (1977) *Outline of a Theory of Practice*, Cambridge: Cambridge University Press.

Bourne, R. (1916) 'Trans-national America', *Atlantic Monthly*, 118 (July): 86–97.

Boyer, C. (1996a) *Cybercities*, New York: Princeton Architectural Press.

Boyer, C. (1996b) *City of Collective Memory*, Cambridge MA: MIT Press.

Bratzel, S. (1999) 'Conditions of success in sustainable urban transport policy—policy change in 'relative successful' European cities', *Transport Reviews*, 19, 2: 177–190.

Brenner, N. and Keil, R. (2006) *The Global Cities Reader*, Abingdon: Routledge.

Brenner, N. and Theodore, N. (eds) (2002) *Spaces of Neoliberalism: Urban Restructuring in North America and Western Europe*, Oxford: Blackwell.

Brenner, N. and N. Theodore (2002) 'Cities and the Geographies of "Actually Existing Neoliberalism"', *Antipode*, 34(3): 349–379.

Breward, C. and Gilbert, D. (eds) (2006) *Fashion's World Cities*, Oxford: Berg. Bridge, G. (2005) *Reason in the City of Difference*, London: Routledge.

Brown, M. (1995) 'Ironies of distance: an ongoing critique of the geographies of AIDS', *Environment and Planning D: Society and Space*, 13(2): 159–183.

Brown, M. (1999) 'Reconceptualizing public and private in urban regime theory: governance in AIDS politics', *International Journal of Urban and Regional Research*, 23(1): 45–69.

Brunhes, J. (1920) *Human Geography*, Chicago: Rand McNally.

Bunnell, T. (1999) 'Views from above and below: the Petronas Twin Towers and/in contesting visions of development in contemporary Malaysia', *Singapore Journal of Tropical Geography*, 20(1): 1–23.

Bunnell, T. (2004a) *Malaysia, Modernity and the Multimedia Super Corridor: A Critical Geography of Intelligent Landscapes*, London: Routledge.

Bunnell, T. (2004b) 'Re-viewing the *Entrapment* controversy: Megaprojection, (mis)representation and postcolonial performance', *GeoJournal*, 59(4): 297–305.

Burgess, E. (1925) 'The growth of the city: An introduction to a research project', in R. Park, E. Burgess and R. McKenzie (eds), *The City,* Chicago: University of Chicago Press.

Buttimer, A. (1978) 'Charisma and context', in D. Ley, and M. Samuels (eds) *Humanistic Geography: Prospects and Problems*, Chicago: Maaroufa Press.

Caldeira, T. (2000) *City of Walls: Crime, Segregation, and Citizenship in Sao Paulo*, Berkeley: University of California Press.

Calvert, C. (2000) *Voyeur Nation: Media, Privacy, and Peering in Modern Culture*, Boulder, CO: Westview Press.

Castells, M. (1989) *The Informational City*, Oxford: Blackwell.

Castells, M. (1997) *The Power of Identity*, Oxford: Blackwell.

Clarke, D.B. (ed.) (1997) *The Cinematic City*, London: Routledge.

Chung, C.J. (ed.) (2002) *The Harvard Design School Guide to Shopping*, Cologne: Taschen.

Cloke, P., Philo, C. and Sadler, D. (1991) *Approaching Human Geography: An Introduction to Contemporary Theoretical Debates,* New York: Guilford.

Coaffee, J. (2004) 'Rings of steel, rings of concrete and rings of confidence: designing out terrorism in Central London pre and post September 11th', *International Journal of Urban and Regional Research*, 28(1): 201–211.

Cody, J.W. (2003) *Exporting American Architecture, 1870–2000*, London: Routledge.

Cocks, C. (2001) *Doing the Town: The Rise of Urban Tourism in the United States, 1850–1915*, Berkeley: University of California Press.

Conradson, D. and Latham, A. (2005) 'Transnational urbanism: attending to everyday practices and mobilities', *Journal of Ethnic and Migration Studies*, 21(2): 227–233.

Conradson, D. and Latham, A. (2007) 'The experiential economy of London: Antipodean transnationals and the overseas experience', *Mobilities*, 2(2): 231–254.

Cook, I. and Harrison, M. (2003) 'Cross over food: re-materializing postcolonial geographies', *Transactions of the Institute of British Geographers*, 28: 297–317.

Couldry, N. (2000) *The Place of Media Power: Pilgrims and Witnesses of the Media Age*, London: Routledge.

Couldry, N. and McCarthy, A. (2004) 'Introduction. Orientations: mapping MediaSpace', in N. Couldry and A. McCarthy (eds), *MediaSpace: Place, Scale and Culture in a Media Age*, London: Routledge. pp. 1–18.

Cox, K. (1993) 'The local and the global in the new urban politics: a critical review', *Environment and Planning D: Society and Space*, 11: 433–448.

Cox, K. (1998) 'Spaces of dependence, spaces of engagement and the politics of scale, or: looking for local politics', *Political Geography*, 17: 1–23.

Crabtree, L. (2006) 'Disintegrated houses: exploring ecofeminist housing and urban design options', *Antipode*, 38(4): 711–734.

Crang, M. and Travlou, P.S. (2001) 'The city and topologies of memory', *Environment and Planning D: Society and Space*, 19(2): 161–177.

Crang, P. and Dwyer, C. (2002) 'Fashioning ethnicities: the commercial spaces of multiculture', *Ethnicities*, 2(3): 410–430.

Crang. P., Dwyer, C. and Jackson, P. (2003) 'Transnationalism and the spaces of commodity culture', *Progress in Human Geography*, 27(4): 438–456.

Crang, M., Graham, S. and Crosbie, T. (2006) 'Variable geometries of connection: urban digital divides and the uses of information technology', *Urban Studies*, 43: 2551–2570.

Crawford, M. (1999) 'Blurring the boundaries: public space and private life', in J. Chase, M. Crawford and J. Kaliski, (eds), *Everyday Urbanism*, New York: Monacelli Press. pp. 22–35.

Cresswell, T. (2006) *On the Move: Mobility in the Modern Western World*, London: Routledge.

Crewe, L. (2001) 'The besieged body: geographies of retailing and consumption', *Progress in Human Geography*, 25(4): 629–640.

Crewe, L. and Gregson, N. (1998) 'Tales of the unexpected: exploring car boot sales as marginal spaces of consumption', *Transactions of the Institute of British Geographers*, 23: 39–53.

Crewe, L. and Lowe, M. (1995) 'Gap on the map? Towards a geography of consumption and identity', *Environment and Planning A*, 27: 1877–1898.

Crilley, D. (1993) 'Megastructures and urban change: aesthetics, ideology and design', in P.L. Knox (ed.), *The Restless Urban Landscape*, Englewood Cliffs, NJ: Prentice Hall. pp. 127–164.

Cronin, A. (2006) 'Advertising and the metabolism of the city: urban space, commodity rhythms', *Environment & Planning D: Society and Space*, 24(4): 615–632.

Cronon, W. (1991) *Nature's Metropolis: Chicago and the Great West,* New York: W.W. Norton.

Crouch, D., Jackson, R. and Thompson, F. (eds) (2005) *The Media and the Tourist Imagination: Converging Cultures,* London: Routledge.

Cunningham, S. (2004) 'The creative industries after cultural policy: a geneology and some possible preferred futures', *International Journal of Cultural Studies,* 7(1): 105–115.

Curtin, M. (2003) 'Media capital: towards the study of spatial flows', *International Journal of Cultural Studies,* 6(2): 202–228.

Darian-Smith, E. (1999) *Bridging Divides: The Channel Tunnel and English Legal Identity in the New Europe,* Berkeley: University of California Press.

Datta, K., McIlwaine, C., Evans, Y., Herbert, J., May, J. and Wills, J. (2007) 'From coping strategies to tactics: London's low-pay economy and migrant labour', *British Journal of Industrial Relations* 45(2): 404–432.

Davis, M. (1990) *City of Quartz: Excavating the Future in Los Angeles,* London: Vintage.

Davis, M. (1992) *Beyond Blade Runner: Urban Control, the Ecology of Fear,* Westview, NJ: Open Media.

Davis, M. (1998) *Ecology of Fear: Los Angeles and the Imagination of Disaster,* New York: Metropolitan Books.

Davis, M. (2000) *Magical Urbanism: Latinos Reinvent the American City,* London: Verso.

Dayan, D. and Katz, E. (1992) *Media Events: The Live Broadcasting of History,* Harvard University Press, Cambridge, Massachusetts.

Dear, M. (1987) 'The postmodern challenge: re-constructing human geography', *Transactions of the Institute of British Geographers,* NS 13: 262–274.

Dear, M. (2000) *The Postmodern Urban Condition,* Oxford: Blackwell.

Dear, M. (2003) 'Superlative urbanisms: the necessity of rhetoric in social theory', *City and Community,* 2(3): 201–204.

Dear, M. (ed.) (2002) *From Chicago to L.A.: Revisioning Urban Theory,* London: Sage.

Dear, M. and Wolch, J. (1987) *Landscapes of Despair: From Deinstitutionalisation to Homelessness,* Princeton: Princeton University Press.

Dear, M., Leclerc, G. and Homero Vill, R. (eds) (1999) *Urban Latino Cultures: La vida latina en LA,* London: Sage.

Debord, G. (1994) *The Society of the Spectacle,* New York: Zone Books.

DeCerteau, M. (1984) *The Practice of Everyday Life,* Berkeley: University of California Press.

Deleuze, G. (1989) *Spinoza: Practical Philosophy,* San Francisco: City Lights Press.

Deleuze, G. (1999) *Foucault,* London: Athlone.

Deleuze, G. (2006) *Two Regimes of Madness: Texts and Interviews 1976–1995,* New York: Semiotexte.

Dickinson, R. (1947) *City Region and Regionalism: A Geographical Contribution to Human Ecology,* London: Kegan Paul.

Dodge, M. and Kitchin, R. (2004) 'Flying through code/space: the real virtuality of air travel', *Environment and Planning A,* 36(2): 195–211.

Dodge, M. and Kitchen, R. (2005) 'Code and the transduction of space', *Annals of the Association of American Geographers,* 95(1): 162–180.

# Bibliography

Dodge, M. and Kitchin, R. (2007) 'The automatic management of drivers and driving spaces', *Geoforum*, 38: 264–275.

Doel, M.A. (1999) *Poststructuralist Geographies: The Diabolical Art of Spatial Science*, Edinburgh: Edinburgh University Press.

Dorling, D., Rigby, J., Wheeler, B., Ballas, D., Thomas, B., Fahmy, E., Gordon, D. and Lupton, R. (2007) *Poverty, Wealth and Place in Britain, 1968–2005*, Bristol: Policy Press.

Dowling, R. (2000) 'Cultures of mothering and car use in suburban Sydney: a preliminary investigation', *Geoforum*, 31: 345–353.

Driver, F. and Gilbert, D. (eds) (1999) *Imperial Cities: Landscape, Display, and Identity*, Manchester: Manchester University Press.

Duncan, O., Cuzzort, R. and Duncan, B. (1961) *Statistical Geography*, Glencoe, Ill.: Free Press.

Duncan, O. and Duncan, B. (1955) 'A methodological analysis of segregation indexes', *American Sociological Review*, 20: 210–217.

Dunn, D. (2005) '"We are not here to make a film about Italy, we *are* here to make a film about ME ...": British television holiday programmes' representations of the tourist destination', in D. Crouch, R. Jackson and F. Thompson (eds), *The Media and the Tourist Imagination: Converging Cultures*, London: Routledge. pp. 154–169.

Dunn, K.M. (2005) 'Repetitive and troubling discourses of nationalism in the local politics of mosque construction in Sydney, Australia', *Environment and Planning D: Society and Space*, 23: 29–50.

Durkheim, E. (1947) *The Division of Labor in Society* (trans. George Simpson), New York: The Free Press.

Duruz, J. (2005) 'Eating at the borders, culinary journeys', *Environment and Planning D*, 23: 51–69.

Edensor, T. (2002) *National Identity, Popular Culture and Everyday Life*, Oxford: Berg.

Ehrenreich, B. and Hochschild, A. (eds) (2003) *Global Woman: Nannies, Maids and Sex Workers in the New Economy*, London: Granta.

Eliot Hurst, M. (ed.) (1974) *Transportation Geography: Comments and Readings*, New York: McGraw-Hill.

Erkip, F. (2003) 'The shopping mall as an emergent urban space in Turkey', *Environment and Planning A*, 35(6): 1073–1093.

Ewen, S. (1999) *All Consuming Images: The Politics of Style in Contemporary Culture*, revised edition, New York: Basic Books.

Fainstein, S. (2001) *The City Builders: Property Development in New York, and London, 1980–2000*, Lawrence, KS: University Press of Kansas.

Featherstone, M. (1991) *Consumer Culture and Postmodernism*, London: Sage.

Featherstone, M., Thrift, N. and Urry, J. (eds) (2005) *Automobilities*, London: Sage.

Fincher, R. and Jacobs J. (eds) (1998) *Cities of Difference*, London: Guilford.

Fischer, C. (1982) *To Dwell Among Friends*, Chicago: University of Chicago Press.

Fishman, R. (1987) *Bourgeois Utopias: The Rise and Fall of Suburbia*, New York: Basic Books.

Florida, R. (2002) *The Rise of the Creative Class*, New York: Basic Books.

Flyvbjerg, B., Bruzelius, N. and Rothengatter, W. (2003) *Megaprojects and Risk: An Anatomy of Ambition,* Cambridge: Cambridge University Press.

Fogelson, R.M. (2001) *Downtown: Its Rise and Fall, 1850–1950,* New Haven: Yale University Press.

Fogelson, R. (2003) *Bourgeois Nightmares: Suburbia, 1870–1930,* New Haven, CT: Yale University Press.

Forest, B. and Johnson, J. (2002) 'Unraveling the threads of history: soviet-era monuments and post-soviet national identity in Moscow', *Annals of the Association of American Geographers,* 92(3): 524–547.

Foucault, M. (1995) *Discipline and Punish,* Translated by A. Sheridan, New York: Random House.

Frazier, J., Margai, F. and Tetteyfio, E. (2003) *Race and Place: Equity Issues In Urban America,* Boulder, CO: Westview Press.

Frieden, B.J. and Sagalyn, L.B. (1989) *Downtown, Inc.: How America Rebuilds Cities,* Cambridge MA: MIT Press.

Friedmann, J. and Wolff, G. (1982) 'World city formation: an agenda for research and action', *International Journal of Urban and Regional Research,* 6(3): 309–344.

Friesen, W., Murphy, L. and Kearns, R. (2005) 'Spiced-up Sandringham: Indian transnationalism and new suburban spaces in Auckland, New Zealand', *Journal of Ethnic and Migration Studies,* 21(2): 227–233.

Fritzsche, P. (1996) *Reading Berlin 1900,* Cambridge, Mass: Harvard University Press.

Fug, G. (1999) *City Making: Building Communities without Building Walls,* Princeton: Princeton University Press.

Fyfe, N. (ed.) (1998) *Images of the Street: Planning, Identity and Control in Public Space,* London: Routledge.

Gandy, M. (1997) 'The making of a regulatory crisis: the restructuring of New York City's water supply', *Transactions of the Institute of British Geographers,* 22(3): 338–358.

Gandy, M. (1999) 'The Paris sewers and the rationalization of urban space', *Transactions of the Institute of British Geographers,* 24: 23–44.

Gandy, M. (2002) *Concrete and Clay: Reworking Nature in New York City,* Cambridge MA: MIT Press.

Gandy, M. (2005a) 'Cyborg urbanization: complexity and monstrosity in the contemporary city', *International Journal of Urban and Regional Research,* 29(1): 26–49.

Gandy, M. (2005b) 'Learning from Lagos', *New Left Review,* 33: 37–52.

Gans, H. (1962) *The Urban Villagers: Group and Class in the Life of Italian-Americans,* New York: Free Press.

Gans, H. (1967) *Levittowners: Ways of Life and Politics in a New Suburban Community,* New York: Pantheon.

Garreau, J. (1991) *Edge City: Life on the New Frontier,* New York: Anchor Books.

Gilbert, D. (2006), 'From Paris to Shanghai: the changing geographies of fashion's world cities'. in C. Breward and G. Gilbert (eds), *Fashion's World Cities,* Oxford: Berg. pp. 3–32.

Glass, R. (1972) 'Anti-urbanism', in M. Stewart (ed.) *The City: Problems of Planning,* London: Penguin. pp. 63–71.

Glennie, P.D. and Thrift, N.J. (1992) 'Modernity, urbanity and mass consumption', *Environment and Planning D: Society and Space,* 10: 423–443.

Glick Shiller, N., Basch, L. and Szanton Blanc, C. (1992) *Towards a Transnational Perspective on Migration: Race, Class, Ethnicity, and Nationalism*, New York: New York Academy of Sciences.

Glick Shiller, N., Basch, L. and Szanton Blanc, C. (1995) 'From immigrant to transmigrant: theorizing transnational migration', *Anthropological Quarterly*, 68(1): 48–63.

Goss, J. (1993) 'The "magic of the mall": an analysis of form, function, and meaning in the contemporary retail built environment', *Annals of the Association of American Geographers*, 83(1): 18–47.

Goss, J. (1999) 'Once-upon-a-time in the commodity world: an unofficial guide to the mall of America', *Annals of the Association of American Geographers*, 89(1): 45–75.

Gottdiener, M. (1994) *Postmodern Semiotics: Material Culture and the Forms of Postmodern Life,* Oxford: Blackwell.

Gottdeiner, M. (1997) *The Theming of America: Dreams, Visions, and Commerical Spaces,* Boulder: Westview Press.

Gottmann, J. (1957) 'Megalopolis or the urbanization of the Northeastern seaboard', *Economic Geography,* 33(3): 189–200.

Gottmann, J. (1961) *Megalopolis: The Urbanized Northeastern Seaboard of the United States*, New York: The Twentieth Century Fund.

Graham, S. (1997) 'Cities in the real-time age: the paradigm challenge of telecommunications to the conception and planning of urban space', *Environment and Planning A*, 29(1): 105–127.

Graham, S. (2000) 'Constructing premium network spaces: reflections on infrastructure networks and contemporary urban development', *International Journal of Urban and Regional Research*, 24(1): 183–200.

Graham, S. (ed.) (2004a) *Cities, War and Terrorism,* Oxford: Blackwell.

Graham, S. (2004b) 'Constructing urbicide by bulldozer in the Occupied Territories', in Graham, S. (ed.) *Cities, War and Terrorism*, Oxford: Blackwell. pp. 192–213.

Graham, S. (2004c) 'Vertical geopolitics: Baghdad and after', *Antipode*, 36(1): 1–23.

Graham, S. and Aurigi, A. (1997) 'Virtual cities, social polarisation and the crisis in urban public space', *Journal of Urban Technology*, 4: 19–52.

Graham, S. and Marvin, S. (1996) *Telecommunications and the City: Electronic Spaces, Urban Places,* New York: Routledge.

Graham, S. and Marvin, S. (2001) *Splintering Urbanism: Networked Infrastructures, Technological Mobilities, and the Urban Condition,* London: Routledge.

Graham, S. and Wood, D. (2003) 'Digitizing surveillance: categorization, space, inequality', *Critical Social Policy,* 75: 227–248.

Greenberg, M. (2003) 'The limits of branding: the World Trade Center, fiscal crisis and the marketing of recovery', *International Journal of Urban and Regional Research*, 27(2): 386–416.

Gregson, N., Metcalfe, A. and Crewe, L. (2007) 'Identity, mobility, and the throwaway society', *Environment and Planning D: Society and Space*, 25(4): 682–700.

Grosz, E. (1994) *Volatile Bodies: Towards a Corporeal Feminism*, Bloomington: University of Indiana Press.

Groth, P. (1994) *Living Downtown: The History of Residential Hotels in the United States,* Berkeley: University of California Press.

203

GUST (Ghent Urban Studies Team) (1999) *The Urban Condition: Space, Community and Self in the Contemporary Metropolis,* Rotterdam: 010 Publishers.

Habermas, J. ([1962] 1989) *The Structural Transformation of the Public Sphere,* Cambridge, MA: MIT Press.

Hägerstrand, T. (1970) 'What about people in regional science?', *Papers, Regional Science Association,* 24: 1–21.

Hägerstrand, T. (1982) 'Diagram, path and project', *Tijdschift voor Economische en Social Geographie,* 73: 323–339.

Haggett, P. (1965) *Locational Analysis in Human Geography,* London: Edward Arnold.

Haldrup, M. and Larson, J. (2006) 'Material culture of tourism', *Leisure Studies,* 25(3): 275–289.

Hall, P. (1966) *The World Cities,* New York: McGraw-Hill.

Hall, P. (1980) *Great Planning Disasters,* London: Weidenfeld and Nicolson.

Hall, P. (1996) *Cities of Tomorrow: An Intellectual History of Urban Planning and Design in the Twentieth Century,* revised edition, Oxford: Blackwell.

Hall, P. (2000) 'Creative cities and economic development', *Urban Studies,* 37(4): 639–649.

Hall, T. (2006) *Urban Geography,* 3rd edition, London: Routledge.

Hall, T. and Hubbard, P. (eds) (1998) *The Entrepreneurial City: Geographies of Politics, Regime and Representation,* Chichester: Wiley.

Hammett, J. and Hammett, K. (eds) (2007) *The Suburbanization of New York,* New York: Princeton University Press.

Hannerz, U. (1996) *Transnational Connections,* London: Routledge.

Hannerz, U. (1998) 'Transnational research', in H.R. Bernard (ed.), *Handbook of Methods in Cultural Anthropology,* Walnut Creek, CA: Altamira Press. pp. 235–256.

Hannigan, J. (1998) *Fantasy City: Pleasure and Profit in the Postmodern Metropolis,* Routledge: London.

Haraway, D. (1991) *Simians, Cyborgs and Women: The Reinvention of Nature,* London: Routledge.

Harris Ali, S. and Keil, R. (2006) 'Global cities and the spread of infectious disease: the case of Severe Respiratory Syndrome (SARS) in Toronto, Canada', *Urban Studies,* 43(3): 491–509.

Harris, C. and Ullman, E. (1945) 'The nature of cities', *The Annals of the American Academy of Political and Social Science,* 242: 7–17.

Harvey, D. (1969) *Explanation in Human Geography,* Arnold: London.

Harvey, D. (1973) *Social Justice and the City,* Baltimore: Johns Hopkins University Press.

Harvey, D. (1982) *The Limits to Capital,* Oxford: Blackwell.

Harvey, D. (1985) *The Urbanization of Capital,* Oxford: Blackwell.

Harvey, D. (1987) 'Flexible accumulation through urbanization: reflections on 'postmodernism' in the American city', *Antipode,* 19(3): 260–286.

Harvey, D. (1989a) 'From managerialism to entrepreneurialism: the transformation of governance in late capitalism', *Geografiska Annaler,* 71B: 3–17.

Harvey, D. (1989b) *The Condition of Postmodernity,* Oxford: Blackwell.

Harvey, D. (1994) 'The invisible political economy of architectural production', in O. Bouman and R. Van Toorn (eds), *The Invisible in Architecture,* London: Academy. pp. 420–427.

Harvey, D. (2000) *Spaces of Hope*, Berkeley: University of California Press.

Harvey, D. (2003) *Paris: Capital of Modernity*, New York: Routledge.

Hebbert, M. (2005) 'The street as locus of collective memory', *Environment and Planning D: Society and Space*, 23(4): 581–596.

Herbert, D. (1972) *Urban Geography: A Social Perspective*, New York: Praeger.

Herod, A. and Aguiar, L.L.M. (2006) 'Introduction: cleaners and the dirty work of neo-liberalism', *Antipode*, 38(3): 425–434.

Herzog, L.A. (2006) *Return to the Center: Culture, Public Space and City Building in a Global Era*, Austin: University of Texas Press.

Hetherington, K. (2004) 'Consumption, disposal, and absent presence', *Environment and Planning D: Society and Space*, 22(1): 157–173.

Hinchliffe, S. (1999) 'Cities and natures: intimate strangers', in J. Allen, D. Massey and M. Pryke (eds), *Unruly Cities?* London: Open University Press.

Hinchliffe, S., Kearns, M., Degen, M. and Whatmore, S. (2005) 'Urban wild things: a cosmopolitical experiment', *Environment and Planning D: Society & Space*, 23(5): 643–658.

Hirsch, A. (1983) *Making the Second Ghetto: Race and Housing in Chicago, 1940–1960*, Chicago: University of Chicago.

Hitchings, R. (2003) 'People, plants and performance: on actor network theory and the material pleasures of the private garden', *Social and Cultural Geography*, 4(1): 99–113.

Hitchings, R. (2007) 'Expertise and inability: cultured materials and the reason for some retreating lawns in London', *Journal of Material Culture*, 11(3): 364–381.

Hollands, R. and Chatterton, P. (2003) 'Producing nightlife in the new urban entertainment economy: corporatisation, branding, and market segmentation', *International Journal of Urban and Regional Research*, 27(2): 223–232.

House of Commons Health Committee 2004, 3rd report of session 2003–4, vol. 1. http://www.publications.parliament.uk/pa/cm200304/cmselect/cmhealth/23/2304.htm

Howard, E. (1996 [1898]) 'Garden cities of tomorrow', 'Author's introduction' and 'The town-country magnet', in *The City Reader*, R. LeGates and F. Stout (eds), London: Routledge. pp. 345–353.

Hoyle, B. and Knowles, R. (1992) *Modern Transport Geography*, London: Belhaven.

Hubbard, P. (2002) 'Screen-shifting: consumption, riskless risks and the changing geographies of cinema', *Environment and Planning A*, 34(4): 1239–1258.

Hubbard, P. (2003) 'Going out/staying in: the seductions of new urban leisure spaces', *Leisure Studies*, 22(3): pp. 255–272.

Hubbard, P. (2006) *The City*, London: Routledge.

Hubbard, P. and Lilley, K. (2004) 'Pacemaking the modern city: the urban politics of speed and slowness', *Environment and Planning D: Society and Space*, 22: 273–294.

Hughes, A. and Reimer, S. (eds) (2004) *Geographies of Commodity Chains*, London: Routledge.

Huxtable, A.L. (1997) *The Unreal America: Architecture and Illusion*, New Press: New York.

Imrie, R. (2003) '"Architects" conceptions of the human body', *Environment and Planning D: Society and Space*, 21(1): 47–65.

Imrie, R. and Raco, M. (eds) (2003) *Urban Renaissance? New Labour, Community, and Urban Policy*, Bristol: Policy Press.

Iveson, K. (2007) *Publics and the City*, Oxford: Blackwell.

Jackson, J.B. (1984) *Discovering the Vernacular Landscape*, London: Yale University Press.

Jackson, J.B. (1994) *A Sense of Place, a Sense of Time*, London: Yale University Press.

Jackson, K. (1985) *Crabgrass Frontier: The Suburbanization of the United States*, New York: Oxford University Press.

Jackson, P. (1984) 'Social disorganisation and moral order in the city', *Transactions of the Institute of British Geographers*, 9: 68–80.

Jackson, P. (1985) 'Urban ethnography', *Progress in Human Geography*, 9: 157–176.

Jackson, P. (2000) 'Rematerialising social and cultural geography', *Social and Cultural Geography*, 1(1): 9–14.

Jackson, P., Lowe, M. and Miller, D. (eds) (2000) *Commercial Cultures: Economies, Practices, Spaces*, Oxford: Berg.

Jackson, P., Crang, P. and Dwyer, C. (eds) (2004) *Transnational Spaces*, London: Routledge.

Jacobs, J. (1961) *The Death and Life of Great American Cities*, New York: Norton.

Jacobs, J.M. (1994a) *Edge of Empire: Postcolonialism and the City*, London: Routledge.

Jacobs, J.M. (1994b) 'Negotiating the heart: heritage, development and identity in postimperial London', *Environment and Planning D: Society and Space*, 12: 751–772.

Jacobs, J.M. (2006) 'A geography of big things', *Cultural Geographies*, 13(1): 1–27.

Jacobs, J.M., Cairns, S. and Strebel, I. (2007) '"A tall storey ... but, a fact just the same": the red road high-rise as a black box', *Urban Studies*, 44(30): 609–629.

Jain, S. (2004) '"Dangerous instrumentality": The bystander as subject in automobility', *Cultural Anthropology*, 19(1): 61–94.

Jargowsky, P. (1997) *Poverty and Place: Ghettoes, Barrios, and the American City*, New York: The Russell Sage Foundation.

Jencks, C. (2005) *The Iconic Building: The Power of Enigma*, London: Frances Lincoln.

Jenkins, H. (2004) 'The cultural logic of media convergence', *International Journal of Cultural Studies*, 7(1): 33–43.

Jessop, B., Peck, J. and Tickell, A. (1999) 'Retooling the machine: economic crisis, state restructuring, and urban politics', in A.E.G. Jonas and D. Wilson (eds), *The Urban Growth Machine: Critical Perspectives, Two Decades Later*, Albany: SUNY Press. pp. 141–159.

Johnson, N.C. (2002) 'Mapping monuments: the shape of public space and cultural identities', *Visual Communication*, 1(3): 293–298.

Johnston, R. (1971) *Urban Residential Patterns: An Introductory Review*, Bell: London.

Johnston, R. (1978) *Multivariate Statistical Analysis in Geography*, Longman: London.

Johnston, R. (1983) *Geography and Geographers: Anglo-American Human Geographers since 1945*, Arnold: London.

Johnston, R., Gregory, D. and Smith, D. (1986) *The Dictionary of Human Geography*, 3rd edition, Oxford: Blackwell.

Jonas, A.E.G. and Wilson, D. (eds) (1999) *The Urban Growth Machine: Critical Perspectives, Two Decades Later*, Albany: SUNY Press.

Judd, D.R. and Swanstrom, T. (2006) *City Politics: the Political Economy of Urban America*, 5th edition, New York: Pearson Longman.

Kaika, M. (2004) *City of Flows: Modernity, Nature and the City,* London: Routledge.

Katz, C. and Kirby, A. (1991) 'In the nature of things: the environment and everyday life', *Transactions of the Institute of British Geographers,* 16(3): 259–271.

Katz, J. (1999) 'Pissed-off in L.A.', in *How Emotions Work,* Chicago: University of Chicago Press.

Katz, J. and Aakhus, M. (eds) (2002) *Perpetual Contact; Mobile Communication, Private Talk, Public Performance,* Cambridge: Cambridge University Press.

Kearney, M. (1995) 'The local and the global: the anthropology of globalization and transnationalism', *Annual Review of Anthropology,* 24: 547-565.

Kearns, M. (2003) 'Geographies that matter – the rhetorical deployment of physicality?', *Social and Cultural Geography,* 4(2): 139–152.

Keeling, D.J. (2007) 'Transportation geography: new directions on well-worn trails', *Progress in Human Geography,* 31(2): 217–225.

Keohane, R. and Nye, J. (eds) (1972) *Transnational Relations and World Politics,* Cambridge MA.: Harvard University Press.

Kesteloot, C., van Weesep, J. and White, P. (eds) (1997) 'Minorities in West European cities', *Tijdschrift voor Economische en Sociale Geografie,* 88: 2.

Kipfer, S. and Keil, R. (2002) 'Toronto Inc.? Planning the competitive city in the New Toronto', *Antipode,* 227–64.

Klein, N. (2000) *No Logo: Taking Aim at the Brand Bullies,* New York: Picador.

Klein, N. (2004) *The Vatican to Vegas: A History of Special Effects,* New York: New Press.

Knauer, L.M. and Walkowitz, D.J. (2004) 'Introduction', in D.J. Walkowitz, and L.M. Knauer (eds), *Memory and the Impact of Political Transformation in Public Space,* Duke University Press, Durham, pp. 1–18.

Knox, P. (1982) *Urban Social Geography: An Introduction,* Harlow: Longman.

Knox, P. (1987) 'The social production of the built environment: architects, architecture and the post-modern city', *Progress in Human Geography,* 11: 354–377.

Knox, P. (1991) 'The restless urban landscape: economic and sociocultural change and the transformation of Metropolitan Washington, DC', *Annals of the Association of American Geographers,* 81(2): 181–209.

Knox, P. and Taylor, P.J. (eds) (1995) *World Cities in a World-System,* Cambridge: Cambridge University Press.

Knox, P. and Pinch, S. (2006) *Urban Social Geography: An Introduction,* 5th Edition, Englewood Cliffs, NJ: Prentice Hall.

Koskela, H. (2000) '"The Gaze without Eyes". Video surveillance and the changing nature of urban space', *Progress in Human Geography,* 24(2): 243–265.

Kotkin, J. (2001) *The New Geography: How the Digital Revolution is Reshaping the American Landscape,* New York: Random House.

Kwan, M.P. (2002) 'Feminist visualisation: re-envisioning GIS as a method in feminist geographic research', *Annals of the Association of American Geographers,* 92: 645–661.

Ladd, A. (1997) *The Ghosts of Berlin: Confronting German History in the Urban Landscape,* Chicago: University of Chicago Press.

Landry, C. and Bianchini, F. (1995) *The Creative City,* London: Demos.

Larner, W. (2007) 'Expatriate experts and globalising governmentalities: the New Zealand diaspora strategy', *Transactions of the Institute of British Geographers,* NS, 32(3): 331–345.

Larsen, J. (2005) 'Families seen sightseeing: performativity of tourist photography', *Space and Culture*, 8(4): 416–434.

Larsen, J., Urry, J. and Axhausen, K. (2006) *Mobilities, Networks, Geographies*, Aldershot: Ashgate.

Latham, A. (2003a) 'Research, performance, and doing human geography: some reflections on the daily-photo diary-interview method', *Environment and Planning A*, 35: 1993–2017.

Latham, A. (2003b) 'Urbanity, lifestyle and the politics of the new urban economy', *Urban Studies*, 40(9): 1699–1724.

Latham, A. (2004) 'American dreams, American empires, American cities', *Urban Geography*, 25 (8): 788–791.

Latham, A. (2006) 'Sociality and the cosmopolitan imagination: national, cosmopolitan and local imaginaries in Auckland, New Zealand', in J. Binny, J. Holloway, S. Millington and C. Young (eds), *Cosmopolitan Urbanism*, London: Routledge.

Latham, A. and McCormack, D.P. (2004) 'Moving cities: rethinking the materialities of urban geographies', *Progress in Human Geography,* 28(6): 701–724.

Latham, A. and McCormack, D. (2008) 'Speed and slowness', in T. Hall, P. Hubbard, and J.R. Short (eds), *The Urban Studies Compendium*, London: Sage.

Latour, B. (1999) *Pandora's Hope: Essays on the Reality of Science Studies*, Cambridge: Harvard University Press.

Latour, B. (2005) *Reassembling the Social: An Introduction to Actor-Network Theory*, Oxford: Oxford University.

Latour, B. and P. Weibel (eds) (2005) *Making Things Public: Atmospheres of Democracy,* Cambridge, MA: MIT Press.

Laurier, E. and Philo, C. (2006a) 'Cold shoulders and napkins handed: gestures of responsibility', *Transactions of the Institute of British Geographers,* 31(2): 193–207.

Laurier, E. and Philo, C. (2006b) 'Possible geographies: a passing encounter in a café', *Area,* 38(4): 353–363.

Law, R. (1999) 'Beyond "women and transport": towards new geographies of gender and urban mobility', *Progress in Human Geography,* 23: 567–588.

Lees, L. (2001), 'Towards a critical geography of architecture: the case of an ersatz Colosseum', *Ecumene*, 8: 51–86.

Lees, L. (2002) 'Rematerialising geography: the 'new' urban geography', *Progress in Human Geography,* 26: 101–112.

Lees, L. (ed.) (2004) *The Emancipatory City: Paradoxes and Possibilities*, London: Sage.

Lefebvre, H. (1991) *The Production of Space*, Oxford: Blackwell.

Lefebvre, H. (2004) *Rhythmanalysis: Space, Time, and Everyday Life.* Edited and translated by S. Elden and G. Moore. London: Continuum.

Leitner, H. (1990) 'Cities in pursuit of economic growth: the local state as entrepreneur', *Political Geography Quarterly*, 9: 146–170.

Leitner, H. and Sheppard, E. (2005) 'Unbounding critical geographic research on cities: the 1990s and beyond', in B. Berry and J. Wheeler (eds), *Urban Geography in America, 1950–2000: Paradigms and Personalities,* London: Routledge. pp. 349–372.

Lemanski, C. (2006a) 'Spaces of exclusivity or connection: linkages between a gated community and its poorer neighbour in a Cape Town master-plan development', *Urban Studies*, 43(2): 397–420.

208

Lemanski, C. (2006b) 'The impact of residential desegregation on social integration: evidence from a South African neighbourhood', *Geoforum*, 37: 417–435.

Lennon, J. and Foley, M. (1999) 'Interpretation of the unimaginable: The US holocaust memorial museum, Washington, D.C. and "dark tourism"', *Journal of Travel Research*, 38: 46–50.

Lennon, J. and Foley, M. (2000) *Dark Tourism: The Attraction of Death and Disaster*, London and New York: Continuum.

Leslie, D. and Reimer, S. (2003) 'Gender, modern design, and home consumption', *Environment and Planning D: Society and Space*, 21: 293–316.

Levitt, P. (1994) 'The social basis for Latino small businesses: the case of Dominican and Puerto Rican entrepreneurs', in E. Melendez and M. Uriarte (eds), *Latino Poverty and Economic Development in Massachusetts*, Boston, MA: University of Massachusetts Press.

Levitt, P. (2001) *Transnational Villagers*, Berkeley: University of California Press.

Lewis, I. (1993) *The City in Slang: New York Life and Popular Speech*, Oxford: Oxford University Press.

Ley, D. (1983) *A Social Geography of the City*, New York: Harper and Row.

Ley, D. (1996) *The New Middle Class and the Remaking of the Central City*, Oxford: Oxford University.

Ley, D. (2004) 'Transnational spaces and everyday lives', *Transactions of the Institute of British Geographers*, 29: 151–164.

Ley, D. and Samuels, M. (eds) (1978) *Humanistic Geography: Prospects and Problems*, Chicago: Maaroufa Press.

Leyshon, A. and Thrift, N. (1997) *Money/Space: Geographies of Monetary Transformation*, London: Routledge.

Lichtenberger, E. (1997) 'Harris and Ullman's "The Nature of Cities": The paper's historical context and impact on further research', *Urban Geography*, 18: 7–14.

Lieberson, S. (1963) *Ethnic Patterns in American Cities*, Glencoe, Ill.: The Free Press.

Lieberson, S. (1980) *A Piece of the Pie: Blacks and White Immigrants Since 1880*, Berkeley: University of California Press.

Liebow, E. (1967) *Tally's Corner*, Boston: Little Brown.

Lindner, L. (1996) *The Reportage of Urban Culture: Robert Park and the Chicago School*, Cambridge: Cambridge University Press.

Llewellyn, M. (2004) '"Urban village" or "white house": envisioned spaces, experienced places, and everyday life at Kensal House, London in the 1930s', *Environment and Planning D: Society and Space*, 22: 229–249.

Longhurst, R. (2005) 'Fat bodies: developing geographical research agendas', *Progress in Human Geography*, 29(3): 247–259.

Low, S. (2004) *Behind the Gates: Life, Security, and the Pursuit of Happiness in Fortress America*, New York: Routledge.

Logan, J.R. and Molotch, H.L. (1987) *Urban Fortunes: the Political Economy of Place*, Berkeley: University of California Press.

Luke, Timothy (1997) 'At the end of nature: cyborgs, "humachines", and environments in postmodernity', *Environment and Planning A: Society and Space*, 29: 1367–1380.

Lukinbeal, C. (1998) 'Reel-to-real urban geographies: the top five cinematic cities in North America', *The California Geographer*, 38: 64–78.

Lyon, D. (2001) *Surveillance Society: Monitoring Everyday Life*, Milton Keynes: Open University Press.

Lyon, D. (ed.) (2003) *Surveillance as Social Sorting: Privacy, Risk and Digital Discrimination*, London: Routledge.

Lyon, D. (2004) 'Technology vs "terrorism": circuits of city surveillance since September 11, 2001', in S. Graham (ed.), *Cities, War and Terrorism: Towards an Urban Geopolitics*, Oxford: Blackwell. pp. 288–311.

McCarthy, A. (2003) *Ambient Television: Visual Culture and Public Space*, Durham: Duke University Press.

McCormack, D. (2005) 'Diagramming practice and performance', *Environment and Planning D: Society and Space*, 23(1): 119–147.

McDowell, L. (1995) 'Body work: heterosexual gender performance in city workplaces', in D. Bell, and G. Valentine (eds), *Mapping Desire: Geographies of Sexuality*, Routledge, London.

McDowell, L. (1997) *Capital Culture: Gender at Work in the City*, Oxford: Blackwell.

McDowell, L. (1999) *Gender, Identity and Place: Understanding Feminist Geographies*, Minneapolis: University of Minnesota Press.

McDowell, L., Batnitzky, A. and Dyer, S. (2007) 'Division, segmentation and interpellation: the embodied labours of migrant workers in a Greater London hotel', *Economic Geography*, 81(1): 1–26.

McGuigan, J. (2004) *Rethinking Cultural Policy*, Milton Keynes: Open University Press.

McKenzie, E. (1994) *Privatopia: Homeowner Associations and the Rise of Residential Private Government*, New Haven: Yale University Press.

McLuhan, M. (1962) *The Gutenberg Galaxy: The Making of Typographic Man*, London: Routledge & Kegan Paul.

McNeill, D. (2001a) 'Barcelona as imagined community: Pasqual Maragall's spaces of engagement', *Transactions of the Institute of British Geographers*, 26: 340–352.

McNeil, D. (2001b) 'Embodying a Europe of the cities: the geographies of mayoral leadership', *Area*, 33: 353–359.

McNeill, D. (2003a) 'Mapping the European urban left: the Barcelona experience'. *Antipode*, 35(1): 74–94.

McNeill, D. (2003b) 'Rome, global city? Church, state and the jubilee 2000', *Political Geography*, 22: 535–556.

McNeill, D. (2005) 'Skyscraper geography', *Progress in Human Geography*, 29(1): 41–55.

McNeill, D. (2007) 'Office buildings and the signature architect: Piano and Foster in Sydney', *Environment and Planning A*, 39: 487–501.

McNeill, D. (2008a) *The Global Architect: Firms, Fame and Urban Form*, New York: Routledge.

McNeill, D. (2008b) 'The hotel and the city', *Progress in Human Geography*, 32(3): 383–398.

McNeill, D., Dowling, R. and Fagan, R. (2005) 'Sydney/global/city: an exploration', *International Journal of Urban and Regional Research*, 29(4): 935–944.

Machimura, T. (1998) 'The urban restructuring process in Tokyo in the 1980s: transforming Tokyo into a world city', *International Journal of Urban and Regional Research*, 16(1): 114–128.

MacLeod, G. (1998) 'In what sense a region? Place hybridity, symbolic shape, and institutional formation in (post-) modern Scotland', *Political Geography*, 17 (7): 833–863.

Mackenzie, A. (2002) *Transductions: Bodies and Machines at Speed*, London: Continuum.

Malbon, B. (1999) *Clubbing: Dancing, Ecstasy and Vitality*, London: Routledge.

Mansvelt, J. (2005) *Geographies of Consumption*, London: Sage.

Marcuse, P. (1997) 'The ghetto of exclusion and the fortified enclave: new patterns in the United States', *American Behavioural Scientist*, 41(3): 311–326.

Marcuse, P. (2004) 'The "war on terrorism" and life in cities', in S. Graham (ed.), *Cities, War and Terrorism: Towards and Urban Geopolitics*, Oxford: Blackwell. pp. 263–275.

Marston, S.A. (2000) 'The social construction of scale', *Progress in Human Geography*, 24(2): 219–242.

Marston, S.A. (2002) 'Making difference: conflict over Irish identity in the New York City St Patrick's Day parade', *Political Geography*, 21: 373–392.

Marvin, S. and Medd, W. (2006) 'Metabolisms of obe*city*: flows of fat through bodies, cities and sewers', *Environment and Planning A*, 38: 313–324.

Massey, D. (2005a) *For Space*, London: Sage.

Massey, D. (2005b) *Strangers in a Strange Land: Humans in an Urbanizing World*, New York: W.W. Norton.

Massey, D. and Denton, N. (1988) 'The dimensions of residential segregation', *Social Forces*, 67(2): 281–315.

Massey, D. and Denton, N. (1993) *American Apartheid: Segregation and the Making of the Underclass*, Cambridge, MA: Harvard University Press.

Massey, D., White, M. and Phua, V.–C. (1996) 'The dimensions of segregation revisited', *Sociological Methods and Research*, 25: 172–206.

Massey, D., Allen, J. and Sarre, P. (eds) (1999) *Human Geography Today*, Cambridge: Polity.

Massumi, B. (2002) *Parables for the Virtual: Movement, Affect, Sensation*, Durham, NC: Duke University Press.

Massumi, B. (2005) 'The future birth of the affective fact', available at http://www. radicalempiricism.org/biotextes/textes/massumi.pdf (accessed March 1st, 2007)

Matless, D. (2000) 'Five objects', in S. Naylor, J. Ryan and I. Cook (eds), *Cultural Turns/Geographical Turns*, London: Pearson.

May, J., Wills, J., Datta, K., Evans, Y., Herbert, J. and McIlwaine, C. (2007) 'Keeping London working: global cities, the British state, and London's new migrant division of labour', *Transactions of the Institute of British Geographers*, 32: 151–167.

May, T. (2005) *Deleuze: An Introduction*, Cambridge: Cambridge University Press.

Mayer, H. and Kohn, C. (1959) *Readings in Urban Geography*, Chicago: Chicago University Press.

Mead, L. (1986) *Beyond Entitlement: The Social Obligations of Citizenship*, New York: The Free Press.

Merrifield, A. (1993a) 'The struggle over place: Redeveloping American can in Southeast Baltimore', *Transactions of the Institute of British Geographers*, 18: 102–121.

Merrifield, A. (1993b) 'The Canary Wharf debacle: from "TINA" – there is no alternative – to "THEMBA" – there must be an alternative', *Environment and Planning A*, 25: 1247–1265.

Merrifield, A. (2002) *Metromarxism: A Marxist Tale of the City*, New York: Routledge.

211

# Bibliography

Merriman, P. (2006) '"A new look at the English landscape": landscape architecture, movement and the aesthetics of motorways in early postwar Britain', *Cultural Geographies,* 13: 78–105.

Merriman, P. (2007) *Driving Spaces: A Cultural-Historical Geography of England's M1 Motorway,* Oxford: Blackwell Publishing.

*Metropolis* (1929) Directed by Fritz Lang.

Meyrowitz, J. (1985) *No Sense of Place: the Impact of Electronic Media on Social Behaviour,* New York: Oxford University Press.

Michael, M. (2006) *Technoscience and Everyday Life: the Complex Simplicities of the Mundane,* Maidenhead: Open University Press.

Miles, M., Hall, T. and Borden, I. (eds) (2003) *The City Cultures Reader,* London: Routledge.

Miller, D. (1989) *Lewis Mumford: A Life,* New York: Weidenfield and Nicolson.

Miller, D. (ed.) (2001) *Car Cultures,* Oxford: Berg.

Miller, D. (ed.) (2005) *Materiality,* Durham, NC: Duke University Press.

Miller, D., Jackson, P., Thrift, N., Holbrook, B. and Rowlands, M. (1998) *Shopping Place and Identity,* London and New York: Routledge.

Mitchell, D. (2003) *The Right to the City: Social Justice and the Fight for Public Space,* New York: The Guilford Press.

Mitchell, K. (1995) 'Flexible circulation in the Pacific Rim: capitalisms in cultural context' *Economic Geography,* 71(4): 364–382.

Mitchell, K. (1997) 'Transnational discourse: bringing geography back in', *Antipode,* 29(2): 101–114.

Mitchell, K. (2003) 'Monuments, memorials, and the politics of memory', *Urban Geography,* 24(5): 442–459.

Mitchell, W. (1995) *City of Bits: Space, Place and the Infobahn,* Cambridge, MA: MIT Press.

Mitchell, W. (1999) *E-Topia: Urban Life Jim But Not as We Know It,* Cambridge, MA: MIT Press.

Mollenkopf, J. and Castells, M. (eds) (1991) *Dual City: Restructuring New York,* New York: Russell Sage.

Monkhouse, F. and Wilkinson, H. (1973) *Maps and Diagrams,* Methuen: London.

Monmonier, M. (1996) *How to Lie with Maps,* 2nd edition, Chicago and London: University of Chicago Press.

Monmonier, M. (2002) *Spying with Maps: Surveillance Technologies and the Future of Privacy,* Chicago: University of Chicago Press.

Morley, D. and Robins, K. (eds) (1996) *Spaces of Identity: Global Media, Electronic Landscapes and Cultural Boundaries,* London: Routledge.

Mullins, P., Natalier, K., Smith, P. and Smeaton, B. (1999) 'Cities and consumption spaces', *Urban Affairs Review,* 35(1): 44–71.

Mumford, L. (1938) *The Culture of Cities,* London: Secker and Warburg.

Murdoch, J. (2006) *Post-Structuralist Geography: A Guide to Relation Space,* London: Sage.

Murray, C. (1984) *Losing Ground: American Social Policy, 1950–1980,* New York: Basic Books.

Neyland, D. (2006) 'The accomplishment of spatial adequacy: analysing CCTV accounts of British town centres', *Environment and Planning D: Society and Space,* 24: 599–613.

# Bibliography

Nye, D.E. (1996) *American Technological Sublime*, Cambridge MA: MIT Press.

O'Neill, P.M. and McGuirk, P. (2003) 'Reconfiguring the CBD: work and discourses of design in Sydney's office space', *Urban Studies*, 40(9): 1751–1767.

O'Sullivan, D. and Wong, D. (2007) 'A surfaced based approach to measuring spatial segregation', *Geographical Analysis*, 39: 147–168.

Olds, K. and Yeung, H. (1999) 'Reshaping "Chinese" business networks in a globalising era', *Environment and Planning D: Society and Space,* 17(5): 535–555.

Olds, K. (1995) 'Globalisation and the production of new urban spaces: Pacific Rim mega-projects in the late 20th century', *Environment and Planning A*, 27: 1713–1743.

Olds, K. (2001) *Globalization and Urban Change: Capital, Culture and Pacific Rim Mega-Projects,* Oxford: Oxford University Press.

Olmstead, F. L. (1996 [1870]) 'Public parks and the enlargement of towns', in R. Legates, and F. Stat (eds), *The City Reader*, London: Routledge. pp. 337–344.

Ong, A. (1999) *Flexible Citizenship: The Cultural Logics of Globalization*, London: Duke University Press.

Pain, R. (2001) 'Gender, race, age and fear in the city', *Urban Studies*, 38: 899–913.

Park, R. (1923) 'The natural history of the newspaper', *The American Journal of Sociology*, 29(3): 273–289.

Park, R. (1925) 'The city: suggestions for the investigation of human behaviour in the urban environment', in R. Park, E. Burgess and R. McKenzie (eds), *The City*, Chicago: University of Chicago Press. pp. 1–46.

Paul, D.E. (2004) 'World cities as hegemonic projects: the politics of global imagineering in Montreal', *Political Geography*, 23(5): 571–596.

Peach, C. (1996) 'Does Britain have ghettos?', *Transactions of the Institute of British Geographers,* 22: 216–235.

Peach, C. (ed.) (1975) *Urban Social Segregation,* London: Longman.

Peach, C., Robinson, V. and Smith, S. (eds) (1981) *Ethnic Segregation in Cities,* London: Croom Helm.

Peck, J. (2005) 'Struggling with the creative class', *International Journal of Urban and Regional Research,* 29(4): 740–770.

Peleman, K. (2002) 'The impact of residential segregation on participation in associations: the case of Moroccan women in Belgium', *Urban Studies*, 39: 727–747.

Peterson, P. (1981) *City Limits,* Chicago: University of Chicago Press.

Philpot, T. (1978) *The Slum and the Ghetto: Neighborhood Deterioration and Middle-Class Reform, Chicago, 1880–1930*, New York: Oxford University Press.

Philo, C. (1995) 'Animals, geography, and the city: notes on inclusions and exclusions', *Environment and Planning D: Society and Space*, 13(6): 655–681.

Pile, S. (1996) *The Body and the City: Psychoanalysis, Subjectivity and Space,* London: Routledge.

Pile, S. (1999) 'What is a city?', in D. Massey, J. Allen, and S. Pile (eds), *City Worlds* London: Routledge. pp. 3–52.

Pinck, P. (2000) 'From the sofa to the crime scene: Skycam, local news and the televisual city', *Urban Space and Representation*, in N. Balshaw, and L. Liam Kennedy (eds), London: Pluto. pp. 55–68.

Pinder, D. (2005) *Visions of the City: Utopianism, Power and Politics in Twentieth-Century Urbanism,* Edinburgh: Edinburgh University Press.

Ponce de Leon, C. (2002) *Self-Exposure: Human Interest Journalism and The Emergence of Celebrity in America 1890–1940*, Chapel Hill: University of North Carolina Press.

Poole, M. and Boal, F. (1973) 'Religious residential segregation in Belfast in mid-1969: a multilevel analysis', *The Institute of British Geographers,* 25: 321–332.

Portes, A. (1996) 'Global villagers: the rise of transnational communities', *The American Prospect,* 2: 74–77.

Portes, A. and Rumbaut, R. (1990) *Immigrant America: A Portrait,* Berkeley: University of California Press.

Portes, A., Guarnizo, L. E. and Landolf, P. (1999) 'The study of transnationalism: the pitfalls and promise of an emergent research field', *Ethnic and Racial Studies,* 22(2): 217–237.

Power, E. (2005), 'Human–nature relations in suburban gardens', *Australian Geographer,* 36: 39–53.

Power, M. and Sidaway, J. (2005) 'Deconstructing twinned towers: Lisbon's Expo '98 and the occluded geographies of discovery', *Social and Cultural Geography,* 6(6): 865–883.

Pred, A. (1966) *The Spatial Dynamics of US Urban-Industrial Growth, 1800–1914: Interpretative and Theoretical Essays,* Cambridge, MA: MIT Press.

Pred, A. (1977) *City-Systems in Advanced Economies: Past Growth, Present Processes and Future Development Options,* London: Hutchinson.

Prince, H. (1980) 'Review of humanistic geography: prospects and problems', by David Ley and Marwyb S. Samuels, *Annals of the Association of American Geographers,* 70(2): 294–296.

Quilley, S. (1999) 'Entrepreneurial Manchester: the genesis of elite consensus', *Antipode,* 31: 185–211.

Quilley, S. (2000) 'Manchester first: from municipal socialism to the entrepreneurial city', *International Journal of Urban and Regional Research,* 24: 601–615.

Raco, M. (2005) 'Sustainable Development, rolled-out neo-liberalism and sustainable communities', *Antipode,* 37(2): 324–346.

Rantisi, N. and Leslie, D. (2006) 'Branding the design metropole: the case of Montréal, Canada', *Area,* 38(4): 364–376.

Rappaport, E. (2000) *Shopping for Pleasure: Women in the Making of London's West End,* Princeton: Princeton University Press.

Ratcliffe, J., Stubbs, M. and Shepherd, M. (2004) *Urban Planning and Real Estate Development,* 2nd edition, London: Spon.

Reichl, A.J. (1999) *Reconstructing Times Square: Politics and Culture in Urban Development,* Lawrence, KA: University Press of Kansas.

Revell, K.D. (2003) *Building Gotham: Civic Culture and Public Policy in New York City, 1898–1938,* Baltimore: Johns Hopkins University Press.

Ritzer, G. (1997) *The McDonaldization of Society,* London: Sage.

Robertson, S. (2007) 'Visions of urban mobility: the Westway, London, England', *Cultural Geographies,* 14: 74–91.

Robinson, J. (1998) 'Spaces of democracy: remapping the apartheid city', *Environment and Planning D: Society and Space,* 16(5): 533–548.

Robinson, J. (2006) *Ordinary Cities: Between Modernity and Development,* London: Routledge.

Rodman, G.B. (1996) *Elvis After Elvis: The Posthumous Career of a Living Legend,* London and New York: Routledge.

Rodrigue, J-P., Comtois, C. and Slack, B. (2006) *The Geography of Transport Systems,* New York: Routledge.

Rose, G. (1993) *Feminism and Geography: The Limits of Geographical Knowledge,* Minneapolis: University of Minnesota Press.

Rose, G. (2007) *Visual Methodologies: An Introduction to the Interpretation of Visual Materials,* 2nd edition, London: Sage.

Ross, K. and Nightingale, V. (2003) *Media and Audiences,* Milton Keynes: Open University Press.

Sachs, W. (1984) *For Love of the Automobile: Looking Back into the History of Our Desires,* Berkeley: University of California.

Sack, R. (1988) 'The consumer's world: place as context', *Annals of the Association of American Geographers,* 78(4): 642–664.

Sadler, S. (1998) *The Situationist City,* London: MIT Press.

Sagalyn, L.B. (2001) *Times Square Roulette: Remaking the City Icon,* Cambridge MA: MIT Press.

Sanders, J. (2001) *Celluloid Skyline: New York and the Movies,* London: Bloomsbury.

Sassen, S. (1988) *The Mobility of Labor and Capital. A Study in International Investment and Labor Flow,* Cambridge: Cambridge University.

Sassen, S. (1991) *The Global City: New York, London, Tokyo,* Princeton NJ: Princeton University Press.

Sassen, S. (1995) 'On concentration and centrality in the global city', in P.L. Knox and P.J. Taylor (eds), *World Cities in a World-Economy,* Cambridge: Cambridge University Press. pp. 63–75.

Sassen, S. (2000) 'The global city: strategic site/new frontier', in E.F. Isin (ed.), *Democracy, Ctizenship and the Global City,* London: Routledge. pp. 48–61.

Sassen, S. (2001) *The Global City: New York, London, Tokyo,* 2nd edition, Princeton NJ: Princeton University Press.

Sauer, C.O. (1925) 'The morphology of Landscape', *University of California Publications in Geography,* 2: 19–54.

Schivelbusch, W. (1986) *The Railway Journey: the Industrialization of Time and Space in the Nineteenth Century,* Berkeley: University of California Press.

Schwartz, V. (1999) *Spectacular Realities: Early Mass Culture in Fin-De Siecle Paris,* Berkeley: University of California.

Schwarzer, M. (2004) *Zoomscape: Architecture in Motion and Meaning,* New York: Princeton Architectural Press.

Scott, A. (1980) *The Urban Land Nexus and the State,* London: Pion.

Scott, A. (ed.) (2002) *Global City-Regions: Trends, Theory, Policy,* New York: Oxford University Press.

Scott, A. (2004) 'Cultural-products industries and urban economic development: prospects for growth and market contestation in global context', *Urban Affairs Review,* 39(4): 461–490.

Scott, H. (2003) 'Cultural turns', in J.S. Duncan, N.C. Johnson, and R.H. Schein (eds), *A Companion to Cultural Geography,* Oxford: Blackwell. pp. 24–37.

Self, R. (2003) *American Babylon: Race and the Struggle for Postwar Oakland,* Princeton: Princeton University Press.

# Bibliography

Sennett, R. (1977) *The Fall of Public Man*, London: Faber and Faber.

Sennett, R. (1994) *Flesh and Stone: The Body and the City in Western Civilisation*, New York: W.W. Norton.

Shatkin, G. (1998) '"Fourth world" cities in the global economy: the case of Phnom Penh, Cambodia', *International Journal of Urban and Regional Research*, 22(3): 378–393.

Shaw, G. and Wheeler, D. (1985) *Statistical Techniques in Geographical Analysis*, New York: John Wiley.

Sheller, M. (2003) 'Mobile publics: beyond the network perspective', *Environment and Planning D: Society and Space*, 22(1): 39–52.

Sheller, M. and Urry, J. (2000) 'The city and the car', *International Journal of Urban and Regional Research*, 24: 737–757.

Sheller, M. and Urry, J. (2003) 'Mobile transformations of "private" and "public" life', *Theory, Culture, and Society*, 20(3): 115–133.

Sheller, M. and Urry, J. (eds) (2004) *Tourism Mobilities: Places to Play, Places in Play*, London: Routledge.

Sheller, M. and Urry, J. (eds) (2006) *Mobile Technologies of the City*, London: Routledge.

Shiel, M. and Fitzmaurice, T. (2001) *Cinema and the City: Film Studies and Urban Societies in a Global Context*, Oxford: Blackwell.

Shields, R. (ed.) (1992) *Lifestyle Shopping: the Subject of Consumption*, London: Routledge.

Shields, R. (2003) *The Virtual*, London: Routledge.

Short, J. (1996) *The Urban Order*, Oxford: Blackwell.

Sibley, D. (1995) *Geographies of Exclusion*, London: Routledge.

Sidaway, J.D. and Power, M. (1995) 'Sociospatial transformation in the "postsocialist" periphery: the case of Maputo, Mozambique', *Environment and Planning A*, 27(9): 1463–1491.

Sidorov, D. (2000) 'National monumentalization and the politics of scale: the resurrections of the cathedral of Christ the Savior in Moscow', *Annals of the Association of American Geographers*, 90(3): 548–572.

Silk, M. and Andrews, D. (2006) 'The fittest city in America', *Journal of Sport & Social Issues*, 30(3): 315–327.

Simmel, G. (1997) 'The metropolis and mental life', in D. Frisby and M. Featherstone (eds), *Simmel or Culture*, London: Sage. pp. 174–185.

Simone, A. (2004) *For the City Yet to Come: Changing African Life in Four Cities*, Durham: Duke University Press.

Simpson, L. (2007) 'Ghettos of the mind: the empirical behaviour of indices of segregation and diversity', *Journal of the Royal Statistical Society*, 170, Part 2: 405–424.

Sinclair, U. (1905) *The Jungle*, Broomhall, PA: Chelsea House.

Sklair, L. (2001) *The Transnational Capitalist Class*, Oxford: Blackwell.

Slater, T. (2006) 'The eviction of critical perspectives from gentrification research', *International Journal of Urban and Regional Research*, 30(4): 737–757.

Smith, M.D. (1996) 'The empire filter back: consumption, production and the politics of Starbucks coffee', *Urban Geography*, 17: 502–524.

Smith, D. (1979) 'Towards a theory of gentrification: a back to the city movement of capital not people', *Journal of the American Planning Association*, 45: 538–548.

Smith, M.P. and Guarnizo, L.E. (eds) (1998) *Transnationalism from Below: Comparative Urban and Community Research*, New Brunswick NJ: Transaction Publishers.

# Bibliography

Smith, M.P. (1994) 'Can you imagine? Transnational migration and the globalization of grassroots politics', *Social Text*, 39: 15–33.

Smith, M.P. (1998) 'The global city: whose social construct is it anyway?' *Urban Affairs Review*, 33(4): 482–488.

Smith, M.P. (2001) *Transnational Urbanism: Locating Globalization*, Oxford: Blackwell.

Smith, M.P. (2002) 'Power in place: retheorizing the local and the global', in J. Eade and C. Mele (eds), *Understanding the City: Contemporary and Future Perspectives*, Oxford: Blackwell. pp. 109–130.

Smith, M.P. (2005a) 'Power in place/places of power: contextualizing transnational research', *City and Society*, 17(1): 5–34.

Smith, M.P. (2005b) 'From context and back again: the uses of transnational urbanism', *City and Society*, 17(1): 81–92.

Smith, M.P. (2005c) 'Transnational urbanism revisited, *Journal of Ethnic and Migration Studies'*, 31(2): 235–244.

Smith, M.P. and Bakker, M. (2005) 'The transnational politics of the Tomato King: meaning and impact, *Global Networks*, 5(2): 129–146.

Smith, N. (1979) 'Towards a theory of gentrification: a back to the city movement by capital not people', *Journal of the American Planning Association*, 45: 538–548.

Smith, N. (1984) *Uneven Development*, Oxford: Blackwell.

Smith, N. (1987) 'Of yuppies and housing: gentrification, social restructuring, and the urban dream', *Environment and Planning D: Society and Space*, 5: 151–172.

Smith, N. (1996) *The New Urban Frontier: Gentrification and the Revanchist City*, London: Routledge.

Smith, N. (2002) 'New globalism, new urbanism: gentrification as global urban strategy', *Antipode*, 34(3): 434–457.

Soja, E.W. (1989) *Postmodern Geographies: The Reassertion of Space in Critical Social Theory*, London: Verso.

Soja, E.W. (1996) *Thirdspace: Journeys to Los Angeles and Other Real-and-Imagined Places*, Oxford: Blackwell.

Soja, E.W. (2000) *Postmetropolis: Critical Studies of Cities and Regions*, Oxford: Blackwell.

Sorkin, M. (ed.) (1992) *Variations on a Theme Park*, New York: Hill and Wang.

Sparke, M. (2000) '"Chunnel visions": unpacking the anticipatory geographies of an anglo-European borderland', *Journal of Borderland Studies*, 15: 187–219.

Spigel, L. (2001) 'Media homes then and now', *International Journal of Cultural Studies*, 4(4): 385–411.

Stacey, M. (1969) 'The myth of community studies, *British Journal of Sociology*, 20(2): 134–147.

Stenning, A. (2003) 'Shaping the economic landscapes of postsocialism? Labour, workplace and community in Nowa Huta, Poland', *Antipode*, 35(4): 761–780.

Stone, C.N. (1989) *Regime Politics: Governing Atlanta 1946–1988*, Lawrence, KS: University Press of Kansas.

Sturken, M. and Cartwright, L. (2001), *Practices of Looking: An Introduction to Visual Culture*, Oxford: Oxford University Press.

Sudjic, D. (1992) *The 100 Mile City*, San Diego: Harcourt Brace.

Sudjic, D. (2005) *The Edifice Complex: How the Rich and Powerful Shape the World*, London: Allen Lane.

Sugrue, T. (1996) *The Origins of the Urban Crisis: Race and Inequality in Postwar Detroit,* Princeton: Princeton University Press.

Sui, D.Z. (2003) 'Musings on the fat city: are obesity and urban forms linked?', *Urban Geography,* 24(1): 75–84.

Sutton, C. (1987) 'The Caribbeanization of New York City and the emergence of transnational social systems', in R. Sutton and E. Chaney (eds), *Caribbean Life in New York City: Sociocultural Dimensions,* New York: Centre for Migration Studies. pp. 15–30.

Swyngedouw, E. (1997) 'Power, nature and the city. The conquest of water and the political ecology of urbanization in Guayaquil, Ecuador: 1880–1980', *Environment and Planning A,* 29(2): 311–332.

Swyngedouw, E. (2007) 'Technonatural relations: the scalar politics of Franco's hydro-social dream for Spain, 1939–1975', *Transactions of the Institute of British Geographers,* 32: 9–28.

Swyngedouw, E., Moulaert, F. and Rodriguez, A. (2002), 'Neoliberal urbanization in Europe: large-scale urban development projects and the new urban policy', *Antipode,* 34: 542–577.

Tauber, K. and Tauber, A. (1965) *Negroes in Cities: Residential Segregation and Neighborhood Change,* Chicago: University of Chicago.

Taylor, G. (1949) *Urban Geography,* London: Methuen.

Thrift, N. (1993) 'For a new regional geography', *Progress in Human Geography,* 17: 92–100.

Thrift, N. (1996) 'New urban eras and old technological fears: reconfiguring the good-will of electronic things', *Urban Studies,* 33(8): 1463–1493.

Thrift, N. (2003) 'Space: the fundamental stuff of human geography', in S. Holloway, S.P. Rice, and G. Valentine (eds), *Key Concepts in Human Geography,* London: Sage.

Thrift, N. (2005a) 'Beyond mediation: three new material registers and their consequences', in D. Miller (ed.), *Materiality.* Durham, NC: Duke University Press. pp. 231–255.

Thrift, N. (2005b) 'But malice aforethought: cities and the natural history of hatred', *Transactions of the Institute of British Geographers,* NS, 30: 133–150.

Thrift, N. (2005c) 'Torsten Hagerstrand and social theory', *Progress in Human Geography,* 29(3): 337–340.

Thrift, N. (2007) 'Immaculate Warfare', in D. Gregory and A. Pred (eds), *Violent Geographies: Fear, Terror, and Political Violence,* London: Routledge.

Thrift, N. and Pred, A. (1981) 'Time geography: a new beginning', *Progress in Human Geography,* 5: 277–286.

Till, K.E. (2005) *The New Berlin: Memory, Politics, Place,* Minneapolis: University of Minnesota Press.

Tölölyan, K. (1991) The nation-state and its others', *Diaspora,* 1(1): 3–7.

Tölölyan, K. (1996) 'Rethinking diaspora(s): stateless power in the transnational world', *Diaspora,* 5(1): 3–36.

Tönnies, F. (1963) *Community and Society,* New York: Harper and Row.

Uitermark, J. (2004a) 'Looking forward by looking back: May Day protests in London and the strategic significance of the urban', *Antipode,* 36(4): 706–727.

Uitermark, J. (2004b) 'The co-optation of squatters in Amsterdam and the emergence of a movement meritocracy: a critical reply to Pruijt', *International Journal of Urban and Regional Research,* 28(3): 687–98.

Ullman, E. (1957) *American Commodity Flow: A Geographical Interpretation of Rail and Water Traffic Based on Principles of Spatial Interchange,* Washington: University of Washington Press.

Underhill, P. (2000) *Why We Buy: The Science of Shopping,* New York: Simon and Schuster.

Underhill, P. (2004) *Call of the Mall: The Geography of Shopping,* New York: Simon and Schuster.

Urry, J. (2000) *Sociology Beyond Societies: Mobilities for the Twenty-First Century,* London: Routledge.

Vertovec, S. (1999) 'Conceiving and researching transnationalism', *Ethnic and Racial Studies,* 22(2): 447–662.

Vertovec, S. (2001) 'Transnationalism and identity', *Journal of Ethnic and Migration Studies,* 27(4): 573–582.

Vidler, A. (2001) 'Diagrams of Utopia', in C. de Zegher, and M. Wigley (eds), *The Activist Drawing: Retracing Situationist Architecture from Constant's New Babylon to Beyond,* Cambridge MA: MIT Press. pp. 83–91.

Wacquant, L. (1999) 'Urban marginality in the coming millennium', *Urban Studies,* 36(10): 1639–1647.

Wacquant, L. (2001) Deadly symbiosis: when ghetto and prison meet and mesh, *Punishment and Society,* 3(1): 95–134.

Wacquant, L. (2004) *Body and Soul: Notebooks of an Apprentice Boxer,* Oxford: Oxford University Press.

Wacquant, L. (2007) *A Comparative Sociology of Advanced Marginality,* Cambridge: Polity Press.

Waitt, G. and Head, L. (2001) 'Postcards and frontier mythologies: sustaining views of the Kimberley as timeless', *Environment and Planning D: Society and Space,* 20: 319–344.

Waldinger, R. (1999) *Still the Promised City? African-Americans and New Immigrants in Post Industrial New York,* Cambridge, MA: Harvard University Press.

Walker, R. (1996) 'Another round of globalization in San Francisco', *Urban Geography,* 17(1): 60–94.

Walkowitz, D.J. and Knauer, L.M. (2004) *Memory and the Impact of Political Transformation in Public Space,* Durham: Duke University Press.

Wark, M. (1994) *Virtual Geography: Living with Global Media Events,* Bloomington and Indianapolis: Indiana University Press.

Waters, J. (2005) 'Transnational family strategies and education in the contemporary Chinese diaspora', *Global Networks,* 5(4): 359–377.

Waters, J. (2006) 'Geographies of cultural capital: education, international migration and family strategies between Hong Kong and Canada', *Transactions of the Institute of British Geographers,* NS, 31(2): 179–192.

Webber, M. (1964) *Explorations into Urban Structure,* Philadelphia: University of Pennsylvania.

Weber, R. (2002) 'Extracting value from the city: neoliberalism and urban redevelopment', *Antipode,* 34(3): 519–540.

Webster, C. (ed.) (2006) *Private Cities: Global and Local Perspectives,* London: Routledge.

Wellman, B. (1979) 'The community question', *American Journal of Sociology,* 84: 1201–1331.

Wellman, B. (ed.) (1998) *Networks in the Global Village: Life in Contemporary Communities,* Boulder: Westview Press.

Wellman, B. and Haythornthwaite, C. (eds) (2002) *The Internet in Everyday Life,* Oxford: Blackwell.

Wellman, B., Carrington, P. and Hall, A. (1988) 'Networks as personal communities', in B. Wellman and S. Berkowitz (eds), *Social Structure: A Network Approach,* Cambridge: Cambridge University Press. pp. 130–184.

Wellman, B. and Leighton, B. (1979) 'Network, neighborhoods and communities', *Urban Affair Quarterly,* 14: 363–390.

Welter, V. (2002) *Biopolis: Patrick Geddes and the City of Life,* Cambridge, MA: MIT Press.

Westerbeck, C. and Meyrowitz, J. (2001) *Bystander: A History of Street Photography,* Boston: Little, Brown and Company.

Whatmore, S. (2002) *Hybrid Geographies: Nature, Culture, Space,* London: Sage.

Wheeler, B., Shaw, M., Mitchell, R. and Dorling, D. (2005) *Life in Britain: using millennial Census data to understand poverty, inequality and place,* Bristol: The Policy Press.

Wheeler, J. (2005) 'Assessing the role of spatial analysis in urban geography in the 1960s', in B.J.L. Berry, and J.O. Wheeler (eds), *Urban Geography in America, 1950–2000: Paradigms and Personalities,* New York: Routledge. pp. 117–128.

Whelan, A., Wrigley, N., Warm, D. and Cannings, E. (2002) 'Life in a "food desert"', *Urban Studies,* 39(11): 2083–2100.

While, A. (2003) 'Locating art worlds: London and the remaking of young British art', *Area,* 35(3): 251–263.

While, A. (2006) 'Modernism vs urban renaissance: negotiating post-war heritage in English city centres', *Urban Studies,* 43(13): 2399–2419.

Whyte, A. and Mackintosh, A. (2003) 'Representational politics in virtual urban places', *Environment and Planning A,* 35: 1607–1627.

Williams, R. (1973) *The Country and the City,* London: Chatto and Windus.

Williams, R. (1976) *Keywords: A Vocabulary of Culture and Society,* London: Flamingo.

Wills, J. (2001) 'Community unionism and trade union renewal in the UK: moving beyond the fragments at last?', *Transactions of the Institute of British Geographers,* 26(4): 465–483.

Wilson, A. (2000) *Complex Spatial Systems: The Modelling Foundations of Regional Analysis,* Harlow: Prentice Hall.

Wilson, D. (2007) *Cities and Race: The New American Black Ghetto,* London: Routledge.

Wilson, E. (1991) *The Sphinx in the City: Urban Design, the Control of Disorder, and Women,* London: Virago.

Wilson, J. (1987) *The Truly Disadvantaged: The Inner City, the Underclass, and Public Policy,* Chicago: University of Chicago Press.

Wolch, J., West, K. and Gaines, T. (1995) 'Trans-species urban theory', *Environment and Planning D: Society and Space,* 13(9): 735–760.

Wolch, J. (2002) 'Anima urbis', *Progress in Human Geography,* 26(6): 721–742.

# Bibliography

Wolch, J. (2007) Green urban worlds', *Annals of the Association of American Geographers,* 97(2): 373–384.

Wollen, P. and Kerr, J. (eds) (2002) *Autopia: Cars and Culture,* London: Reaktion.

Women and Geography Study Group of the IBG (1984) *Geography and Gender: An Introduction to Feminist Geography,* Oxford: Blackwell.

Wood, D.M., Lyon, D. and Abe, K. (2007) 'Surveillance in urban Japan: a critical introduction', *Urban Studies,* 44(3): 551–568.

Wrigley, N. and Lowe, M. (2002), *Reading Retail: A Geographical Perspective on Retailing and Consumption Spaces,* London: Arnold.

Wu, F. (ed.) (2005) *Globalization and the Chinese City,* London: Routledge.

Yeung, H. (2005) 'Rethinking relational economic geography', *Transactions of the Institute of British Geographers,* NS, 30: 37–51.

Young, M. and Willmott, P. (1957) *Family and Kinship in East London,* London: Penguin Books.

Zukin, S. (1992) 'The city as a landscape of power: London and New York as global financial capitals', in L. Budd, and S. Whimster (eds), *Global Finance and Urban Living,* Andover: Routledge, Chapman and Hall. pp. 195–223.

Zukin, S. (1995) *The Cultures of Cities,* Cambridge MA: Blackwell.

# AUTHOR INDEX

222

**223**

**224**

# SUBJECT INDEX

*Note*: Page numbers in **bold** indicate a more comprehensive treatment

# Subject Index

**231**

Subject Index